Image Processing and

The Practical Approach Series

SERIES EDITOR

B. D. HAMES
Department of Biochemistry and Molecular Biology
University of Leeds, Leeds LS2 9JT, UK

See also the Practical Approach web site at **http://www.oup.co.uk/PAS**
★ **indicates new and forthcoming titles**

Affinity Chromatography
Affinity Separations
Anaerobic Microbiology
Animal Cell Culture (2nd edition)
Animal Virus Pathogenesis
Antibodies I and II
Antibody Engineering
Antisense Technology
★ Apoptosis
Applied Microbial Physiology
Basic Cell Culture
Behavioural Neuroscience
Bioenergetics
Biological Data Analysis
Biomechanics – Materials
Biomechanics – Structures and Systems
Biosensors
★ C Elegans
Carbohydrate Analysis (2nd edition)
Cell-Cell Interactions
The Cell Cycle
Cell Growth and Apoptosis
★ Cell Growth, Differentiation and Senescence
★ Cell Separation
Cellular Calcium
Cellular Interactions in Development
Cellular Neurobiology
Chromatin
★ Chromosome Structural Analysis
Clinical Immunology
Complement
★ Crystallization of Nucleic Acids and Proteins (2nd edition)
Cytokines (2nd edition)
The Cytoskeleton
Diagnostic Molecular Pathology I and II
DNA and Protein Sequence Analysis
DNA Cloning 1: Core Techniques (2nd edition)
DNA Cloning 2: Expression Systems (2nd edition)

- DNA Cloning 3: Complex Genomes (2nd edition)
- DNA Cloning 4: Mammalian Systems (2nd edition)
- ★ DNA Microarrays
- ★ DNA Viruses
- Drosophila (2nd edition)
- Electron Microscopy in Biology
- Electron Microscopy in Molecular Biology
- Electrophysiology
- Enzyme Assays
- Epithelial Cell Culture
- Essential Developmental Biology
- Essential Molecular Biology I and II
- ★ Eukaryotic DNA Replication
- Experimental Neuroanatomy
- Extracellular Matrix
- Flow Cytometry (2nd edition)
- Free Radicals
- Gas Chromatography
- Gel Electrophoresis of Nucleic Acids (2nd edition)
- ★ Gel Electrophoresis of Proteins (3rd edition)
- Gene Probes 1 and 2
- ★ Gene Targeting (2nd edition)
- Gene Transcription
- Genome Mapping
- Glycobiology
- Growth Factors and Receptors
- Haemopoiesis
- ★ High Resolution Chromotography
- Histocompatibility Testing
- HIV Volumes 1 and 2
- ★ HPLC of Macromolecules (2nd edition)
- Human Cytogenetics I and II (2nd edition)
- Human Genetic Disease Analysis
- ★ Immobilized Biomolecules in Analysis
- Immunochemistry 1
- Immunochemistry 2
- Immunocytochemistry
- ★ *In Situ* Hybridization (2nd edition)
- Iodinated Density Gradient Media
- Ion Channels
- ★ Light Microscopy (2nd edition)
- Lipid Modification of Proteins
- Lipoprotein Analysis
- Liposomes
- Mammalian Cell Biotechnology
- Medical Parasitology
- Medical Virology
- MHC Volumes 1 and 2
- ★ Molecular Genetic Analysis of Populations (2nd edition)
- Molecular Genetics of Yeast
- Molecular Imaging in Neuroscience
- Molecular Neurobiology
- Molecular Plant Pathology I and II
- Molecular Virology
- Monitoring Neuronal Activity

- ★ Mouse Genetics and Transgenics
- Mutagenicity Testing
- Mutation Detection
- Neural Cell Culture
- Neural Transplantation
- Neurochemistry (2nd edition)
- Neuronal Cell Lines
- NMR of Biological Macromolecules
- Non-isotopic Methods in Molecular Biology
- Nucleic Acid Hybridisation
- ★ Nuclear Receptors
- Oligonucleotides and Analogues
- Oligonucleotide Synthesis
- PCR 1
- PCR 2
- ★ PCR 3: PCR In Situ Hybridization
- Peptide Antigens
- Photosynthesis: Energy Transduction
- Plant Cell Biology
- Plant Cell Culture (2nd edition)
- Plant Molecular Biology
- Plasmids (2nd edition)
- Platelets
- Postimplantation Mammalian Embryos
- ★ Post-translational Processing
- Preparative Centrifugation
- Protein Blotting
- ★ Protein Expression
- Protein Engineering
- Protein Function (2nd edition)
- Protein Phosphorylation (2nd edition)
- Protein Purification Applications
- Protein Purification Methods
- Protein Sequencing
- Protein Structure (2nd edition)
- Protein Structure Prediction
- Protein Targeting
- Proteolytic Enzymes
- Pulsed Field Gel Electrophoresis
- RNA Processing I and II
- RNA-Protein Interactions
- Signalling by Inositides
- ★ Signal Transduction (2nd edition)
- Subcellular Fractionation
- Signal Transduction
- ★ Transcription Factors (2nd edition)
- Tumour Immunobiology
- ★ Virus Culture

Image Processing and Analysis

A Practical Approach

Edited by

RICHARD BALDOCK
*MRC Human Genetics Unit,
Western General Hospital,
Crewe Road, Edinburgh EM4 2XU*

and

JIM GRAHAM
*University of Manchester,
Imaging Science and Biomedical Engineering
Stopford Building, Oxford Road,
Manchester M13 9PT*

OXFORD
UNIVERSITY PRESS

Great Clarendon Street, Oxford OX2 6DP

Oxford University Press is a department of the University of Oxford
and furthers the University's aim of excellence in research, scholarship,
and education by publishing worldwide in

Oxford New York

Athens Auckland Bangkok Bogotá Buenos Aires Calcutta
Cape Town Chennai Dar es Salaam Delhi Florence Hong Kong Istanbul
Karachi Kuala Lumpur Madrid Melbourne Mexico City Mumbai
Nairobi Paris São Paulo Singapore Taipei Tokyo Toronto Warsaw
and associated companies in Berlin Ibadan

Oxford is a registered trade mark of Oxford University Press

Published in the United States
by Oxford University Press Inc., New York

© Oxford University Press, 2000

All rights reserved. No part of this publication may be reproduced,
stored in a retrieval system, or transmitted, in any form or by any means,
without the prior permission in writing of Oxford University Press.
Within the UK, exceptions are allowed in respect of any fair dealing for the
purpose of research or private study, or criticism or review, as permitted
under the Copyright, Designs and Patents Act, 1988, or in the case
of reprographic reproduction in accordance with the terms of licences
issued by the Copyright Licensing Agency. Enquiries concerning
reproduction outside those terms and in other countries should be
sent to the Rights Department, Oxford University Press,
at the address above.

This book is sold subject to the condition that it shall not, by way
of trade or otherwise, be lent, re-sold, hired out, or otherwise circulated
without the publisher's prior consent in any form of binding or cover
other than that in which it is published and without a similar condition
including this condition being imposed on the subsequent purchaser

Users of books in the Practical Approach Series are advised that prudent
laboratory safety procedures should be followed at all times. Oxford
University Press makes no representation, express or implied, in respect of
the accuracy of the material set forth in books in this series and cannot
accept any legal responsibility or liability for any errors or omissions
that may be made.

A catalogue record for this book is available from the British Library

Library of Congress Cataloging in Publication Data
(Data available)

ISBN 0-19-963701-6 (Hbk)
0-19-963700-8 (Pbk)

Typeset by Footnote Graphics,
Warminster, Wilts
Printed in Great Britain by Information Press, Ltd,
Eynsham, Oxon.

Preface

Image processing, image analysis, and pattern recognition are techniques now widely used in bioscience and medicine. There is a plethora of packages and systems which one can buy, or download free, ranging from menu-driven systems to libraries for C-programming. There is also a large number of textbooks, with one or more of the above terms in their titles, which explain the computational basis of these techniques. Many of these textbooks are excellent in their mathematical and computational descriptions, but take as their audience engineers and computer scientists for whom the methods themselves present the scientific interest. This volume, in common with others in the *Practical Approach* series, takes a different perspective. We take our audience to be scientists whose interests are in other fields, and for whom these methods provide useful analytical tools. As is the case with all tools it is best for the user to have an idea how they work, so that their behaviour can be understood, without necessarily becoming immersed in the details. As this volume forms part of a series aimed at biologists, we have retained that emphasis. However, the methods we describe are equally applicable in many other fields—the earth sciences for example—and we hope it will be of interest to scientists in these areas also. Try to imagine your favourite images in the place of the chromosomes or fungal mycelia.

As in any field of engineering, there are carefully worked out computational and mathematical principles in image processing and analysis, from which practitioners have developed a collection of rules of thumb and established common procedures. It is the latter aspect, rather than the former, that we hope to emphasize here. Some of these procedures take the form of well-known algorithms, and these, together with rules of thumb and standard processes, are presented as Protocols. Some standard pieces of knowledge are more readily expressed in mathematical formulae. This is a mathematical topic, and it would be impossible to describe it without the use of mathematical language. We understand, however, that mathematics does not hold a deep fascination for at least some of our target audience, and we have tried to keep the mathematical descriptions at a level that would be acceptable to a final-year high-school or first-year university student. Many of the algorithms and formulae are already implemented in software packages and libraries. We hope that the descriptions here will give the user some understanding of the reasoning behind the functions given and the limitations of their operation. The algorithms should be presented in sufficient detail that a competent programmer, without much image analysis experience, should be able to implement them if required.

There is a wide range of material in the image processing and analysis literature. We have deliberately set out to avoid producing another all-encompassing

Preface

book. The material we have chosen has two aims. First, it is intended to cover the basic methods of which a user of the technology should be aware. Second, we present a selection of more advanced techniques that we hope will inspire the application of image analysis to a wider range of complex scientific problems. Each chapter describes in some detail a specific technique or collection of related techniques. To focus on the practicalities, each chapter relates its technical content to one (or maybe two) applications in bioscience or medicine. We hope that the reader should get an appreciation of not only how individual methods can solve real analytical problems, but also the range of applications which can be addressed.

The terms *image processing*, *image analysis*, and *pattern recognition* are often used interchangeably. In fact, they refer to different activities, but they overlap and in a given application, it is likely that all three will be used. Chapters 2, 3, and 4 address each of these topics. There is some overlap in their descriptions, where the individual topics merge. The chapters can be read individually, but together they should give a good grounding of the basic methods for dealing with digital images. These are preceded by a chapter on image acquisition. Many users of image analysis acquire their images using a television camera mounted on a microscope. Both of these components are becoming increasingly complex, and an understanding of their properties and limitations separately and in combination is important for achieving the highest quality data. Later chapters deal with more advanced computational methods, in the use of explicit mathematical models of image appearance, and the analysis of three-dimensional data. Not only in microscopy, but also in other research and clinical fields, analysis of structures in three dimensions is of increasing importance.

Of course there are omissions. Some readers may feel that we might have included material on confocal microscopy or stereology, for example. These are both large topics which we felt are dealt with very well elsewhere for the practising biologist. We have, however, included the rather less well-known topic of projective stereology, which extends the established stereology literature.

We have had guiding principles in the choice of topics, but ultimately those selected, and the illustrative applications, reflect the interests of the editors and the social circles in which they move. We hope you like them.

September 1999

Jim Graham
Richard Baldock

Contents

List of Contributors   xvii
Abbreviations   xix

1. Microscope image acquisition   1
Stephanie L. Ellenberger and Ian T. Young

  1. Introduction   1
      Microscope systems   1
      Case study   2

  2. Basic optics   3
      Nature of light   3
      Illumination   7
      Optical system   9

  3. Microscope image acquisition system   12
      Microscope types and components   12
      Camera types and performance   19
      Images   28

  4. Application   29
      Objective   29
      Materials and methods   30
      Performance   34

  References   35

2. Biological image processing and enhancement   37
G. W. Horgan, C. A. Reid, and C. A. Glasbey

  1. Introduction   37

  2. Contrast manipulation   39
      Functional transformation   40
      Histogram-based transformations   41
      Pseudocolour   42
      Thresholding   43
      Zooming   46

  3. Filtering   46
      Linear filters   46
      Non-linear filters   50

4. Mathematical morphology	53
Basic ideas	54
Greylevel morphology	56
Complex operations	58
Top-hat transform and background subtraction	58
5. Texture measures	60
Statistical approaches	61
Structural approaches	64
Modelling approaches	64
Texture transforms	65
Acknowledgements	66
References	66

3. Image analysis: quantitative interpretation of chromosome images 69

Kenneth R. Castleman

1. Introduction	69
The karotyping problem	70
Historical perspective	70
Current practice	70
Automatic karyotyping	70
2. Image segmentation	72
Thresholding	73
The watershed algorithm	76
Gradient image thresholding	76
Laplacian edge detection	77
Edge detection and linking	78
Region growing	84
3. Boundary refinement	85
Active contours	85
Binary image processing	87
Morphological image processing	88
Boundary curvature analysis	91
Touch and overlap resolution	92
4. Chromosome measurement	93
Morphological features	94
Banding pattern features	94
5. Chromosome classification	96
The Bayes classifier	96
6. Karotype generation	99
Chromosome straightening	99

Chromosome enhancement	99
Chromosome assignment	101
7. Other cytogenetics applications	102
Multiplex-FISH	103
Metaphase finding	104
8. Image processing software	105
Academic packages	105
Commercial packages	105
References	106

4. Pattern recognition: classification of chromosomes — 111

Jim Graham

1. Introduction	111
2. The chromosome classification problem	112
3. Classification methods	114
Defining classification in terms of probabilities	114
Using Bayes' formula	115
Non-parametric methods	121
4. Features and feature selection	124
Selecting features	126
Combining features	128
Clustering	129
5. Neural networks	131
Introduction	131
Supervised training	132
Unsupervised training	135
6. Classifying chromosomes	137
Features	137
Data sets	138
Classifiers	139
7. Classifier validation	141
The need for validation	141
Cross-validation, the jackknife and the bootstrap	141
The confusion matrix	142
Validating issues for neural networks	143
Training-set size and validation	143
Validation issues for the two-class problem	144
8. Available material	148
Software	148
Further reading	149

Contents

Acknowledgements	150
References	150

5. Three-dimensional (3D) reconstruction from serial sections — 153
Fons J. Verbeek

1. Introduction — 153
- 3D reconstruction in microscopy — 153
- Why serial sectioning is necessary — 154

2. Methodological aspects — 156
- Modes of action — 158
- Organizing the reconstruction data — 159
- Methods and devices for the input of the data — 161
- Spatial resolution — 166
- Alignment and deformation correction — 166

3. Mathematical aspects — 173
- Estimation of transformations — 173
- Spatial transformation of image values — 174
- Image correspondences — 175
- Point-pattern matching methods — 175
- Shape-based assessment methods — 178
- Registration and congruencing using image moment — 180
- Affine transform component estimation — 181
- Evaluation using similarity measures — 184
- Intensity-based methods — 185

4. Systems for routine application of 3D reconstruction — 186
- Reconstruction under rigid transformation — 187
- Reconstruction including deformation correction — 187

References — 194

6. 3D analysis: registration of biomedical images — 197
Daniel Rueckert and David J. Hawkes

1. Introduction — 197
- Intra-subject registration — 198
- Inter-subject registration — 198
- Serial registration — 198
- Image to physical space registration — 198
- Overview — 199

Contents

2. Registration transformation — 200
  Rigid transformation — 200
  Affine transformation — 200
  Projective transformation — 201
  Elastic or fluid transformation — 201

3. Registration basis — 202
  Point-based registration — 202
  Contour- and surface-based registration — 204
  Voxel-based registration — 207

4. Optimization — 211

5. Applications — 212
  2D registration — 212
  3D–3D rigid registration — 212
  3D–3D non-rigid registration — 214
  Validation — 215

6. Summary — 217

Acknowledgements — 218

References — 219

7. Model-based methods in analysis of biomedical images — 223

Tim Cootes

1. Introduction — 223

2. Background — 224

3. Application — 225

4. Theoretical background — 227
  Building models — 227
  Image interpretation with models — 232
  Active shape models — 234

5. Discussion — 240

6. Implementation — 242

Appendices — 242
  Aligning the training set — 242
  Principal component analysis — 244
  Applying a PCA when there are fewer samples than dimensions — 245
  Aligning two shapes — 246

Acknowledgements — 246

References — 246

8. Projective stereology in biological microscopy — 249
Andrew D. Carothers

1. Introduction — 249
2. Points and distances — 250
3. Lines — 253
4. Surface areas — 254
5. Volumes — 255
6. Applications — 256
   - Determining the order of three DNA loci — 256
   - Location of chromosomal domains in cell nuclei — 257
   - Size of chromosomal domains in cell nuclei — 259
7. Discussion — 259

Acknowledgements — 260

References — 260

9. Image warping and spatial data mapping — 261
Richard A. Baldock and Bill Hill

1. Introduction — 261
   - Notation and numerical methods — 262
2. Image comparison — 264
   - Binary image overlay — 264
   - Colour comparator — 267
   - Blink comparator — 268
3. Image re-sampling — 269
   - Nearest neighbour interpolation — 270
   - Bilinear interpolation — 270
   - Computational efficiency — 270
4. Defining correspondence — 273
5. Global transforms — 274
   - Affine — 275
   - Polynomial — 278
   - Conformal — 279
   - Thin-plate spline — 282
   - Multiquadric — 284
   - Gaussian — 284
6. Local transforms — 285
   - Simple mesh transformation — 285
   - Elastic plate warping — 286

7. Mapping gene-expression data	286
8. Summary	286
References	288

List of Suppliers 289

Index 294

Contributors

RICHARD A. BALDOCK
MRC Human Genetics Unit, Western General Hospital, Crewe Road, Edinburgh EH4 2XU, UK. Richard.Baldock@hgu.mrc.ac.uk

ANDREW D. CAROTHERS
MRC Human Genetics Unit, Western General Hospital, Crewe Road, Edinburgh EH4 2XU, UK. Andrew.Carothers@hgu.mrc.ac.uk

KENNETH R. CASTLEMAN
Perceptive Scientific Instruments, Inc., League City, Texas, USA. Castleman@persci.com

TIM COOTES
University of Manchester, Imaging Science and Biomedical Engineering, Stopford Building, Oxford Road, Manchester M13 9PT, UK. t.cootes@man.ac.uk

STEPHANIE L. ELLENBERGER
Image Processing, Information Systems Division, P.O. Box 155, 2600 AD Delft, The Netherlands. eberger@tpd.tno.nl

C. A. GLASBEY
Biomathematics and Statistics Scotland, The King's Buildings, Edinburgh EH9 3JZ, UK. c.glasbey@bioss.sari.ac.uk

JIM GRAHAM
University of Manchester, Imaging Science and Biomedical Engineering, Stopford Building, Oxford Road, Manchester M13 9PT, UK. Jim.Graham@man.ac.uk

DAVID J. HAWKES
UMDS Radiological Sciences, Guy's Hospital, London SE1 9RT, UK. D.Hawkes@umds.ac.uk

BILL HILL
MRC Human Genetics Unit, Western General Hospital, Crewe Road, Edinburgh EH4 2XU, UK. Bill.Hill@hgu.mrc.ac.uk

G. W. HORGAN
Biomathematics and Statistics Scotland, Rowett Research Institute, Aberdeen AB21 9SB, UK. g.horgan@bioss.sari.ac.uk

C. A. REID
Biomathematics and Statistics Scotland, Rowett Research Institute, Aberdeen AB21 9SB, UK. c.reid@bioss.sari.ac.uk

Contributors

DANIEL RUECKERT
UMDS Radiological Sciences, Guy's Hospital, London SE1 9RT, UK. D.Rueckert@umds.ac.uk

FONS J. VERBEEK
NIOB Hubrecht Laboratory, Uppsalalaan 8, 3584 CT Utrecht, The Netherlands. verbeek@niob.knaw.nl

IAN T. YOUNG
Department of Applied Physics, Delft University of Technology, PO Box 5046, 2600 GA Delft, The Netherlands. young@ph.tu.tudelft.nl

Abbreviations

ADC	analogue-to-digital converter
ADU	analogue-to-digital converter unit
ASM	active shape model
AUC	area under the curve
CAR	computer-assisted reconstruction
CCD	charge-coupled device
CGH	comparative genomic hybridization
CLSM	confocal laser scanning microscope
CRT	cathode-ray tube
CT	computed tomography
CV	coefficient of variation
DAPI	4′,6-diamidino-2-phenylindole
DOF	depth of field
EM	expectation maximization
FEM	finite-element model
FFD	free form deformations
FFT	fast Fourier transform
FISH	fluorescent *in situ* hybridization
FITC	fluorescein isothiocyanate
FN	false negative
FOV	field of view
FP	false positive
FROC	free-response operating characteristic
GIS	geographic information system
ICP	iterative closest point
IR	infrared
ISCN	international system of human cytogenetic nomenclature
kbp	kilobase pairs
k-NN	k-nearest neighbour
LoG	Laplacian of Gaussian
LUT	look-up table
M-FISH	multicolour FISH
MLP	multilayer perceptron
MR	magnetic resonance
MRASM	multiresolution active shape model
NA	numerical aperture
NCC	normalized cross-correlation
NMR	nuclear magnetic resonance
PCA	principal components analysis

Abbreviations

PET	positron emission tomography
PSF	point spread function
RBF	radial basis function
RGB	red green blue
ROC	receiver operating characteristic
ROI	region of interest
RMS	root mean square
SD	sampling density *also* standard deviation
SNR	signal-to-noise ratio
SPECT	single-photon emission computed tomography
SSD	squared sum of differences
SVD	singular value decomposition
TN	true negative
TP	true positive
TRITC	tetramethylrhodamine-5-isothiocyanate
TVP	total vertical projection
UV	ultraviolet
VIR	variance of intensity ratio
WDD	weight density distribution

1

Microscope image acquisition

STEPHANIE L. ELLENBERGER and IAN T. YOUNG

1. Introduction

Modern microscopes allow the attachment of a camera. Technology is available to send acquired images directly to a computer for further processing. In these images object features such as length, area, or intensity can be measured with appropriate software. The measurement results can be used to classify objects or to determine the quality of an object. The automatic determination of object properties is called *automated quantitative microscopy*.

Progress in medicine and industry has stimulated the development of fast and accurate automated microscope systems which process a great number of objects in a short time, for example to monitor the treatment of cancer patients or to inspect a large number of printed circuit boards. Another advantage of automated systems besides speed and accuracy is that results are reproducible if the set-up is unchanged. It is also easy to store results electronically for further processing at a later time or for comparison with other results.

For both image analysis and image processing, attention has to be paid to errors occurring during the image formation and acquisition. For image analysis, which has an image as input and data as output, the quality may not always be critical. Some measurements might be done on noisy images or on unsharp images without losing much accuracy. For image processing, however, the image quality has to be high as the output is an image too. But in both cases we can be sure that the effort invested in getting high quality images is rewarded with less trouble in subsequent procedures.

There are limitations, however, with respect to the accuracy of the object representation by a microscope image. The optics which form an image of the object on the sensor and the digitization process which happens in the camera introduce errors or suppress parts of the object signal. Some of the limitations encountered in image acquisition can be compensated and/or eliminated.

1.1 Microscope systems

Automated microscopy only needs a few components. Besides a microscope, a motorized, computer-controlled stage, a computer-controlled camera with frame grabber, and a computer are required. In special cases (e.g. fluorescence microscopy) computer-controlled filter wheels may be necessary as well. All components are commercially available and some companies already deliver complete systems with application-dependent, ready-to-use software such as the Q550 CW Cytogenetics Workstation by Leica, UK, or the LSC, Laser Scanning Cytometer by CompuCyte, Cambridge, MA, USA. As it is desirable to get information about many different properties of a variety of objects, automated systems can be built around all kinds of microscopes such as brightfield, two-photon, fluorescence, or confocal microscopes.

The advantages of quantitative imaging in microscopy were first pointed out in the 1920s when Hartridge and Roughton published a paper on determining the ratio of oxy- to carboxyhaemoglobin in small blood samples (1). They stressed that the accuracy of measurements could be increased by using photographic methods. Another advantage of using photography was that the data were recorded and could be reviewed at any time.

A further step towards quantitative microscopy was performed by Caspersson in the 1930s (1). He and his collaborators improved and extended both instrumentation and methods for photometric measurements. The image of the specimen was no longer observed with the human eye but with special equipment. Thus results did not depend on human estimation but were obtained with instruments.

Analytical or quantitative microscopy became more and more important as the number of applications for cytophotometry increased and the instruments became more sophisticated which improved the precision of measurements. With the help of accurate photometric methods more insight was achieved into topics such as the relation of chromosomal changes in cells and tumour diagnosis. This gave rise to demands for automated screening in clinical cytology (1).

1.2 Case study

This chapter describes the possibilities and problems of microscope image acquisition. A practical example which is used here to illustrate theoretical considerations is the automated spot counter (2–4). The spot counter is used to determine statistical properties of many interphase cell nuclei which are either stained for brightfield microscopy or for fluorescent microscopy. In the cell nuclei special parts of a chromosome are labelled and appear as dots within the counterstained nuclei. The number of dots per nucleus gives information about the state of a cell. Usually each healthy human cell has 22 autosomes which appear in pairs and two sex chromosomes, either XX (female) or XY (male), for a total of 46 chromosomes per cell nuclei. This spot counter can, for each chromosome type, detect genetic deviations from this standard.

2. Basic optics

2.1 Nature of light

The image formation in a microscope is determined by light. An object may, for example, absorb, transmit, or emit light depending on the biological and biophysical processes that are involved. This light is registered by a sensor and forms the image. Several problems that may occur during image acquisition can be explained by the physical properties of light.

2.1.1 Light as waves

One way to describe light is as an electromagnetic wave. The wave has a certain amplitude which is related to the intensity of the light and a certain wavelength which determines the colour we perceive (*Figure 1*). The human eye can perceive waves with a wavelength between about 380 and 730 nm. The sensitivity of the eye is maximal for wavelengths of about 550 nm which corresponds to green light.

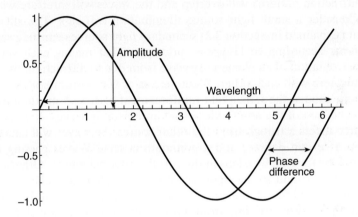

Figure 1. Characterization of waves by wavelength, amplitude, and phase difference.

One observation which can be explained with the wave nature of light is diffraction. If a light beam passes a pinhole in an opaque surface and is imaged by an aberration-free lens, the resulting image is not a point but a concentric circular pattern of alternating dark and bright rings known as the Airy disc (*Figure 2*). The waves seem to bend and extend into the shadowy area. The shape of the Airy disc depends on the wavelength λ of light and the numerical aperture NA (opening) of the objective lens. The smaller the wavelength and the higher the numerical aperture, the smaller the Airy disc. In Section 2.3 we will see how this phenomenon affects the resolution of an optical system.

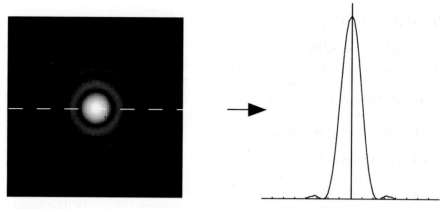

Figure 2. The Airy disc and its profile.

Diffraction occurs at any edge, not only at pinholes. In objects with many edges diffraction patterns will overlap and the waves will interfere with each other. Consider a small light source illuminating two slits. The slits act as coherent (explained in Section 2.2) secondary light sources emitting spherical wave fronts according to Huygens' principle. That means each slit is the centre of excitation of an elementary wave front due to diffraction. Both wave fronts interfere with each other. Where crests of waves from both slits arrive at the same time at the same point reinforcement occurs. A maximum is seen which is the sum of the amplitudes of both interfering waves. If a crest from one source meets a trough from the other source the waves will cancel each other (to a certain degree) and a minimum is seen. Waves passing the slit unbent will be focused by a lens in the zeroth-order maximum. Other maxima follow to the left and to the right (see *Figure 3*). They are called first-order maximum, second-order maximum, etc. Here the path-length difference between the waves arriving from both sources is one wavelength, two wavelengths, etc. To get a sharp image of an object at least the zeroth- and first-order maxima have to pass the objective lens. The better the light collecting power of an objective (high NA), the more maxima are involved in the image formation and the sharper the image will be.

2.1.2 Light as particles

Some properties of light can be explained if we consider light to consist of massless particles, photons. Consider imaging an object two or more times under exactly the same conditions. The images look similar or even identical but a comparison of the single pixel values will show that they differ. Even with perfect optics and modern cameras of high sensitivity it is impossible to get exactly the same images. Global features like average intensity will be the same but the single pixel values will vary from image to image. This is because

1: Microscope image acquisition

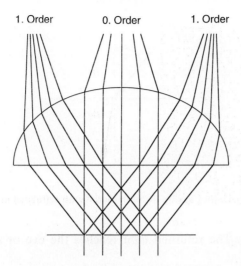

Figure 3. Diffraction at a grating, showing the zeroth- and both first-order maxima.

the number of photons emitted by the light source is subject to quantum physics laws and known to obey Poisson statistics (5, 6). This behaviour limits the quality of the images that can be obtained.

2.1.3 Interaction of light and matter

The image perceived with a microscope is formed by light that interacts with an object placed on the stage. Besides the direct effects discussed above, other effects based on the wave or particle nature of light are:

Reflection and *scatter*—Light that hits an object is returned by the object surface. The direction of the returned light depends on the structure of the object surface. For smooth surfaces such as mirrors, specular reflection occurs. The incident light beam and the reflected light beam form the same angle with the surface (*Figure 4*). This kind of light does not contain much information about the surface from which it is reflected. We only see a bright spot. However, rough surfaces tend to scatter the light in

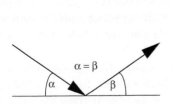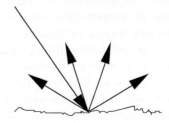

Figure 4. Reflection and scatter.

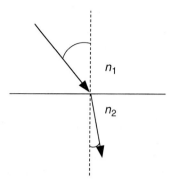

Figure 5. Refraction of light passing through media with refractive indices n_1 and n_2.

all directions. The returned light reaches the eye or any other optical sensor (7, 8).

Refraction—Light beams pass through a medium at a certain speed. The speed of the light is related to the refractive index of the medium. If a light beam changes from one medium to another with a different refractive index, its speed is changed. The beam deviates from a straight line as shown in *Figure 5*. For example, if a stick is dipped into water it will appear to be bent at the air–water interface. But actually the rays are *refracted* at the transition from air to water (7, 8).

Flare—also called glare. Images can be disturbed by light that is reflected from a lens surface, not from the object surface to be observed. This light is unwanted as it can reduce image contrast and form hot spots. To prevent flare it is advisable not to illuminate more than the field of view. Good objectives are provided with antireflection coatings of the optical surfaces and the lens curvature is properly designed in order to hinder flare. (8)

Absorption—White light passing through an object can change in colour and intensity. The object selectively absorbs parts of the white light. More precisely, it decreases the wave amplitude over a range of wavelengths. Transparent objects do not have this property. They do not change the wave amplitude but they may change the wave phase. Transparent objects may be viewed either by making the phase changes visible (as in phase microscopy) or by treating the objects with dyes that cause wavelength-specific light absorption. The treatment with dyes may have the disadvantage that the organisms are killed.

Fluorescence—This is a property of *all* materials. Irradiation with light of a certain wavelength will cause the object to emit light of a different wavelength. The material absorbs a part of the energy from the excitation light which brings the molecules into a higher but unstable energy state. A part of this energy is released when the molecule returns to the energy ground state. As some energy is lost before emission, light of lower energy and thus longer wavelength (other colour) is returned. The difference between the excitation and the emission wavelength is called the Stokes' shift.

2.2 Illumination

In microscopy there may be several light sources which illuminate the specimen in different ways. The differences lie in the phase relationship and wavelength composition of the electromagnetic waves emitted by the light source. Light sources that emit light at a single wavelength or a very narrow band of wavelengths are called *monochromatic* or narrow-band light sources. The most well-known example of a monochromatic source is the laser. Halogen and mercury lamps are broad-band light sources. Laser beams also illuminate objects *coherently*. The emitted electromagnetic waves have a systematic phase relationship. Neighbouring point sources in the laser are not independent of each other. In other light sources such as halogen lamps the phase relationship is random. The illumination is incoherent. In this case the illumination can be changed to partially coherent by closing the aperture. In the limit when the aperture is closed to a pinhole, the illumination is quasi coherent (9).

In confocal brightfield microscopy the illumination is always coherent due to the point detector. However, in fluorescence (confocal) microscopy fluorescent particles are incoherent light sources. Even if illuminated with coherent light the image formation is determined by the emitted fluorescence which is incoherent (10).

The properties of incoherent and coherent illumination are quite different and both have their advantages and disadvantages. For example, a sharp edge imaged in coherent illumination will show ringing effects and the edge will be slightly shifted into the bright area. This must be taken into account when measurements are to be performed on that image. Coherent illumination also introduces the 'speckle effect', which means that the resulting image looks granular. Another important point about coherent illumination is that high image quality can only be obtained with aberration-free optics. Even small dust particles on the lens can produce disturbances in the image. However, the resolution achievable with coherent illumination can be better than with incoherent illumination depending on the phase changes induced by the object (11).

An advantage of incoherent illumination is that under the same conditions more light is present compared to coherent illumination. Furthermore, the image intensity does not change with a change of the focal position.

A straightforward way to control illumination was described by August Köhler in 1893. The aim of his procedure for controlled specimen illumination is the precise illumination of the pupils in the optical path of a microscope. This ensures high resolution and homogeneous illumination.

If the specimen is illuminated directly by a lamp, the filament will cause disturbances in the image due to spatial variation in intensity. By placing a lens in front of the lamp all light is collected and projected into the condenser aperture. The lens has the effect that every point of the light source illuminates the entire field. An aperture diaphragm is used to control the diameter of the illuminated field. It is desirable to limit the illumination to the field of view as otherwise glare might occur or fluorescent particles which are not within the field of view might bleach before they can be examined. The ray bundle produced by the lamp collector lens is focused on the specimen with the condenser. The opening of the condenser aperture determines whether the illumination is coherent, partially coherent, or incoherent. If the aperture of the condenser is larger than or equal to the aperture of the objective, the illumination is incoherent; otherwise it is partially coherent or even coherent if the condenser aperture is close to a point source.

In fact, two separate real planes are found in the microscope—the object image plane and the filament image plane of the lamp. Several planes exist in the optical path where either the specimen or the lamp is projected sharply. These planes are called conjugate planes (*Figure 6*). Each series of conjugated planes determines a different aspect of the imaging performance of a microscope. The first series of planes (planes A_C in *Figure 6*) determine the resolving power and the contrast in the image. By changing the diameter of the aperture, diaphragm resolution and contrast are changed. The other planes (planes B_C in *Figure 6*) determine the diameter of the field of view and the illumination. Changing the diameter of the field diaphragm changes the diameter of the field of view and the illumination.

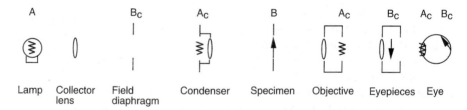

Figure 6. Conjugate planes: the planes A_C are conjugate to the plane A. The planes B_C are conjugate to the plane B.

Protocol 1. Köhler illumination for conventional wide-field microscopes

1. Turn on the light source.
2. Open the field diaphragm as far as possible.
3. Open the condenser diaphragm as far as possible.

4. Adjust the position of the condenser. Its distance to the specimen should be 1–3 mm.
5. Adjust the interpupillary distance of the eyepieces. If the light intensity is too high, reduce the brightness of the lamp.
6. Focus the specimen by moving the stage carefully up and down until the sample is focused. The image may be of poor quality as the illumination is not correct yet.
7. Close the field diaphragm. By moving the condenser up and down focus the image of the field diaphragm. Be careful that you do not lift the sample with the condenser. If you cannot find the edge of the field diaphragm, adjust the position of the condenser by turning the centring screws.
8. Centre the image of the field diaphragm by turning the centring screws.
9. Open the field diaphragm until the field of view is illuminated completely (not further).
10. Remove one eyepiece and look at the back focal plane of the objective. Open the condenser diaphragm until its image nearly fills the exit pupil of the objective.
11. Replace the eyepiece and adjust the light intensity to a comfortable viewing level.

2.3 Optical system

2.3.1 Specifying lenses

A lens can be characterized by its magnification and its numerical aperture (NA). The latter is a measure of the light-collecting power of a lens and helps to determine the resolution. The NA is determined by the refractive index of the immersion medium and the half-angle of acceptance α of the lens:

$$NA = n \sin \alpha \qquad [1]$$

For example, the objectives 25×/0.6 and 40×/1.3 Oil have an NA of 0.6 and 1.3 respectively. The refractive index n for the first objective is 1 for air. For the other objective n is 1.52 for oil. Thus α is about 37° and 59° respectively.

A single lens can magnify an image about 10 to 25 times. Higher magnification is obtained by more complex arrangements of lenses as in (compound) microscope objectives. But here, too, the resolution is limited. Because of the wavelength of light, which is used to form an image, and the diffraction phenomenon, the maximal resolution is about 0.17 μm (see *Equation 2*). By using other illuminators with shorter wavelength (UV-light) and special microscopes, the resolution can be further increased. Note that the resolution does not automatically improve with higher magnification. In order to have a

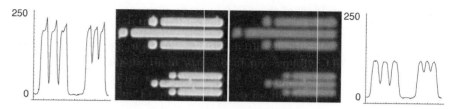

Figure 7. Images of a test pattern and profiles along one line in the image (shown in white). Both images are taken with a 100× Oil objective with variable NA. For the left image the NA is 1.3, for the right image the NA is 0.6. The left image is sharper than the right image. This is confirmed by the corresponding profiles. The profile of the left image has steeper edges than the profile of the right image. The asymmetric peaks in the left profile are due to the partially coherent illumination. As the condenser of the microscope has an NA < 1.3 it is not possible to get perfect Köhler illumination.

higher resolution the numerical aperture has to increase, as can be seen in *Figure 7*, or the wavelength has to decrease.

The performance of a microscope is not described only by the magnification and the numerical aperture of the objective. A much better way to describe the complete imaging system is to determine the point spread function (PSF). This is the image of a point source, known as the Airy disc (see Section 2.1.1). If the image of a point source is known, it is possible to predict the image of any object imaged with the system. The image of a point source gives information about the brightness that is transmitted through the optics, the deformation introduced by the optics, and the maximal possible resolution. Lord Rayleigh defined the resolution as the distance from the centre of the Airy disc to the first dark ring which occurs at approximately:

$$d = 0.61 \frac{\lambda}{\text{NA}} \qquad [2]$$

This definition is arbitrary but well accepted in the field of microscopy. For the example objectives 25×/0.6 and 40×/1.3 Oil, the first dark ring occurs at 0.51 and 0.23 µm respectively (see *Figure 8*). The best possible resolution is obtained at the shortest visible wavelength, about 0.4 µm, and the highest possible NA, about 1.4, which results in $d \sim 0.17$ µm.

It is possible to get a sharp image up to a certain distance from the focus position. This distance is called the depth of field (DOF). It depends on the wavelength λ of light, the numerical aperture NA of the objective lens and the refractive index *n* of the medium between the objective lens and the object.

$$\Delta z_{\text{DOF}} = \pm \frac{n\lambda}{2(\text{NA})^2} \qquad [3]$$

With increasing NA the depth of field decreases. When observing the image through the eyepieces it is much easier to get a sharp image than when observing the specimen with a camera. Unlike the human eye, a camera lens

1: Microscope image acquisition

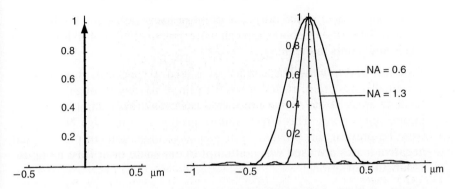

Figure 8. The image of a single point source (left profile) as a normalized Airy disc or a point spread function (right profile). The Airy disc is depicted for two different objectives at a wavelength of 0.5 μm. The objectives are 25×/0.6 and 40×/1.3 Oil.

cannot accommodate, which makes it more difficult to focus the specimen in automated imaging.

The DOF is an indication of the ability to observe three-dimensional structures with a given optical system. The smaller the DOF, the less suitable a lens is to image specimens that extend in the axial direction. Typical examples for the DOF of objectives at a wavelength of 500 nm are shown in *Figure 9*.

2.3.2 Optical distortions

Lenses enable the inspection of small objects which cannot be resolved by the eye. But it is the lens itself which also limits microscopy because of several possible aberrations:

Chromatic aberration—The refractive index of glass differs for different wavelengths. It is larger for blue light (short wave ~ 400 nm) than for red

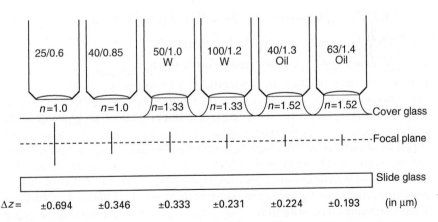

Figure 9. Depth of field for different objectives in μm. Assumed wavelength, 0.5 μm. Indices are $n = 1.0$ (air), 1.33 (water) and 1.52 (oil).

light (longer wave ~ 600 nm). The point of focus varies and the image is blurred. This kind of aberration can cause colour fringes or halos around edges and points in the image.

Spherical aberration—The surface of a lens is spherical. Rays passing marginal zones are refracted more than those passing through the intermediate areas. Axial rays are refracted least. The rays cannot be focused in one single point. This effect is more pronounced for high NA objectives. To prevent this phenomenon special combinations of positive and negative lenses are used which attempt to eliminate or at least minimize spherical aberration.

Coma—A blurred image occurs for non-axial points as marginally passing rays meet in another point than axially passing rays. Usually, microscope objectives are corrected for this kind of aberration.

Astigmatism—The phenomenon that rays of a non-axial point do not have a focus.

Curvature of field—The image is not projected onto a plane parallel to the lens but onto the surface of a sphere with a radius equal to the focal length.

Geometric distortion—The magnification of the lens varies from the centre to the periphery of the field. This causes barrel-formed (greater magnification at centre) or pin-cushion-formed (greater magnification at periphery) distortions.

Objectives, condensers, and eyepieces, corrected for several aberrations, are available (see Section 3.1.3).

3. Microscope image acquisition system

3.1 Microscope types and components

3.1.1 Microscope types and image formation

This chapter is limited to the description of conventional, fluorescence, and confocal microscopy. However, the principles and methods described for these three microscope types can easily be utilized for other types of microscopes as well.

The conventional microscope is a well-thought-out combination of lenses and aperture diaphragms with which it is possible to study details with different light transmission properties. If the object structures do not differ in absorption, dyes can be used to make structures visible. This is often sufficient to get the required information.

In many cases it is possible to label objects or even only specific parts of objects with *fluorochromes*. By using appropriate excitation and emission filters the image is formed by the emitted fluorescence only and not by the illuminating light source. The detection of the specimen detail does not depend on the size of the detail but on the amount of fluorochrome attached to it. This method is very sensitive and results in high contrast images. Fluorescence is not only suitable to detect the presence of certain substances: the emitted light also contains information about the specimen which allows quantitative as well as qualitative analysis.

The disadvantage of both methods is that light originating from out-of-focus planes is also collected in the imaging device. This light disturbs the image and also limits the resolution. A method developed to prevent out-of-focus light from reaching the detector is the confocal microscope. The main features are a point source to illuminate the object and a point detector which only detects in-focus light. In order to get an image of an object, the object has to be scanned point by point. By scanning different planes the three-dimensional structure of an object can be very accurately imaged. The object can be viewed in confocal absorption, confocal reflection or in confocal fluorescence, although the last two are the most common.

In *Figures 10* and *11* the principle of image formation is depicted for some methods. The path of the light is depicted for upright microscopes, but inverted microscopes are possible as well.

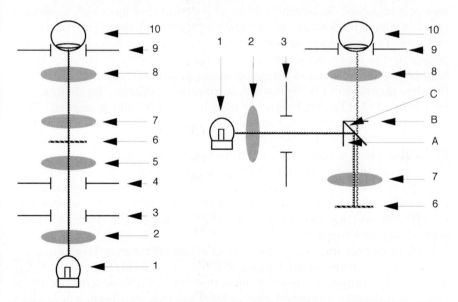

Figure 10. Light path through a conventional microscope for brightfield imaging (left) and a conventional microscope for epi-illumination fluorescence microscopy (right). 1, lamp; 2, collector lens; 3, field diaphragm; 4, condenser diaphragm; 5, condenser lens; 6, specimen; 7, objective lens; 8, eyepieces; 9, eye pupil; 10, eye; A, excitation filter; B, emission filter; C, dichroic mirror.

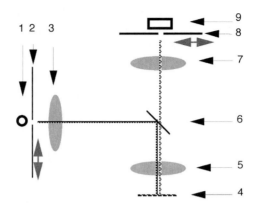

Figure 11. Light path through a fluorescence confocal microscope. 1, point source; 2, pinhole; 3, collector lens; 4, specimen; 5, objective; 6, chromatic reflector; 7, collector lens; 8, pinhole; 9, detector. The point source and the detector are scanned in order to image the whole specimen.

3.1.2 Light sources

Several kinds of lamps can be used to illuminate an object. Depending on the type of microscope the demands differ. It is desirable to have high luminance over the complete spectrum and constant light emission during the lifetime of the lamp. The last condition is not fulfilled by tungsten lamps. Tungsten condenses on the inner surface of the bulb, thus blackening it. Halogen lamps have higher luminance and possess constant brightness for constant voltage. But they do not emit UV-light which is frequently necessary for fluorescence microscopy (12, 13). Arc lamps are used in case UV-light is needed. Their handling is more complicated as they produce a significant amount of heat and some lamps (e.g. xenon) also produce ozone. Arc lamps are generally ignited using high voltage. This requires an electronic starter.

In general, tungsten or halogen lamps are used if the object is observed in absorption mode. They have continuous (broad-band) energy spectra. The power of irradiance reaches its maximum in the infrared part of the spectrum ($\sim$ 800 nm and higher) which heats the specimen. The absolute power depends on the wattage.

For fluorescence microscopy mercury or xenon arc lamps or lasers are used. The energy spectrums of arc lamps can differ considerably. The main difference from the tungsten filament lamp is that arc lamps show strong peaks at several wavelengths. In general, they are brighter than the tungsten or halogen lamps especially at shorter wavelengths. Xenon arc lamps are especially suitable for excitation in the UV. The lifetime and the cost of mercury and xenon arc lamps differ but the costs per running hour are comparable with a slight advantage for xenon arc lamps.

1: Microscope image acquisition

Lasers can be used if (coherent) illumination at a specific wavelength is needed. The laser wavelength determines which fluorophores can be excited. Different lasers are available for different wavelengths (9, 12, 13).

3.1.3 Objectives

The first magnifying lens system in a microscope is the objective. It collects the light emerging from the specimen and forms an inverted and magnified image of the specimen in the intermediate image plane. A magnification up to 100 times is possible. There are several types of objectives for different applications with special optical corrections. Depending on the correction for aberrations they are divided into three groups:

Achromats—These are corrected for chromatic aberrations for two wavelengths (in general, 656 and 486 nm) and spherically corrected for green light. These objectives are less expensive than others as the cost of the material used is relatively low and the correction for chromatic aberrations is less complicated.

Fluorite or *semi-apochromats*—In this kind of objective elements made of special material are used to get a better correction for chromatic aberrations. Achromat lenses consist of crown and flint glass. For fluorite objectives the crown glass is partially or completely replaced by fluorspar or newer, specially formulated, optical glasses (7). Besides better correction for chromatic aberration, this type of objective can have larger numerical apertures than achromats. The light collection power and thus the resolution improves. This kind of objective is intended for fluorescence microscopy as UV-light is transmitted, which is important as several dyes have to be excited with UV. Furthermore, the material used shows little or no autofluorescence.

Apochromats—These are the best but also the most expensive type of objectives. They are corrected for chromatic aberrations for at least three wavelengths (red, 656 nm; green, 546 nm; and blue, 486 nm) and are free from spherical aberrations for two colours. High contrast and high resolution can be obtained with these objectives. However, the material used and the manufacturing are more sophisticated than for the other types resulting in higher prices.

If, in addition, the lens is corrected for curvature of field this is marked on the tube by the inscription 'Plan' or 'Plano'. For imaging it is important to have a field that is sharp in the centre and at the edge. To achieve this the objective has to be corrected accordingly.

Another categorization of lenses is based on the medium between the objective lens and the specimen. This is indicated by the inscription 'Oel' or 'Oil' or 'Water' or 'Glyz'. This indicates that the objective is designed for use with an immersion medium such as oil, water, or glycerine. These media have

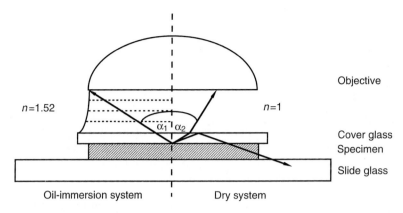

Figure 12. Oil-immersion and dry systems.

a higher refractive index than air which increases the light-collecting power of the lens (see *Equation 1*). At a transition from one medium to another light is bent and reflected (see *Figure 12*). The larger the difference in refractive index the stronger the bending of the light rays. In the case of an oil-immersion system the refractive index is matched perfectly to the refractive index of the coverslip glass and the objective lens ($n = 1.52$). The rays pass without being bent and in addition no reflections occur at the transition from one element to the other. In the case of water-immersion systems ($n = 1.33$) or dry systems ($n = 1.0$) the light rays are refracted at the transition from the embedding medium to the coverslip, from the coverslip to the immersion medium (air or water respectively) and at the transition from the immersion medium to the objective lens. Furthermore, light rays may be reflected at any of the three transitions which will lead to loss of light in the detector plane.

The numerical aperture of objectives making use of other immersion media than air can be larger than 1 with a maximum that is close to the refractive index of the medium as shown in *Table 1* (14–16).

Table 1. Refractive index of common immersion media and the maximum possible numerical aperture (NA) of corresponding immersion objectives

Medium	Refr. index	Max. NA
Air	1.00	0.95
Water	1.33	1.2
Glycerine	1.44	1.3
Oil	1.52	1.4

3.1.4 Other components

Other components of a microscope which affect the optical system are:

Eyepieces—The second optical system that magnifies the specimen comprises the eyepieces. The image presented by the objective in the intermediate image plane is magnified between 5 and 20 times. The eyepieces complete the correction for lens aberrations. Therefore, objective and eyepieces should be from the same manufacturer as the manner of correction can vary. For good results it is vital to ensure that the magnification factor of the objective is larger than that of the eyepieces.

Condenser—A lens system between object and illuminator, the condenser concentrates light onto the object. It focuses an image of the field diaphragm onto the object plane. The effect of a properly focused condenser (according to Köhler) is that the intensity is uniform over the entire illuminated field. The numerical aperture of the illuminating cone of light can be adjusted by the aperture diaphragm to match the NA of the objective (incoherent illumination). As lenses are used, aberrations due to the condenser can also occur. Different types of condensers corrected for various aberrations are available. Excellent quality is achieved with aplanatic-achromatic condensers which are corrected for spherical and chromatic aberrations. It is also possible to provide special types of condensers for other microscopy methods such as immersion or darkfield. For good results it is vital to position the condenser in the centre of the field of view.

Aperture diaphragm—A diaphragm is a hole with a variable diameter. The aperture diaphragm is situated in front of the focal plane of the condenser. It is used to control the convergence of the cone of light emerging from the condenser. Resolution, contrast, and depth of focus are regulated by adjusting the diameter of the diaphragm (17). The aperture of the illumination can be smaller than the aperture of the objective. By closing the aperture diaphragm to two-thirds of the illuminated field the contrast can be improved but the illumination is no longer incoherent and the resolution will decrease. Further closing will not lead to higher contrast.

Field diaphragm—The field diaphragm is used to regulate the field of illumination. Stopping down the field does not affect the numerical aperture so the resolution will be unchanged. It is not advisable to illuminate more of the field of view than necessary as flare (see Section 2.1.3) might occur.

Optical filter—Several filters are used in microscopy for different purposes. The filters can be separated into different groups:

contrast filter—used for absorption objects to suppress or enhance contrast and to select certain parts of the spectrum. Two different filters belong to this group:

colour filter—made from dyed gelatine or bulk-dyed glass. It is available in many colours. Light of the same colour as the filter is transmitted while the complementary colour is absorbed. This type of filter is typically a wide-band filter with either a short-pass (blue-passes shorter wavelengths), long-pass (red-passes longer wavelengths) or bandpass character. If filter and object have the same colour, fine details can be studied. If filter and object have different colours, the contrast is increased.

interference filter—a narrow-band filter. Very small bands ($\Delta\lambda = 20$ nm) of certain wavelengths can be selected with the help of this filter. Because of their sharp optical characteristics and high quality they are expensive.

conversion filter—may be necessary in photomicrography if the light source does not match the response of the film in use.

UV-absorbing filter—a long-pass filter to suppress UV rays which may damage the eyes, the specimen, and the microscope objectives.

heat absorbing or *infrared filter*—absorbs infrared (IR) radiation which causes most of the heating in the optical system. This may introduce a greenish colour cast which can be corrected by a colour correction filter. This kind of filter is important if a charge-coupled device (CCD) camera is used to obtain images. The CCD-sensor is very sensitive to IR light. Without a filter of this kind the signal would be dominated by IR radiation which is not visible to the human eye.

neutral density or *grey filter*—used to lower the light intensity. It is available in many different shades of grey for gradual reduction of brightness without changing the colour.

barrier or *emission filter*—narrow-band wavelength filter which is used in fluorescence microscopy to prevent the remaining excitation light from reaching the eye or camera.

excitation filter—used in fluorescence microscopy to select the part of the spectrum that induces the fluorescence. It is a bandpass filter with a usual width of 10–20 nm.

dichroic filter—used for epi-illuminated fluorescence microscopes. It is an interference filter which separates the excitation light from the fluorescence. In general, the emission light passes through the dichroic filter while the excitation light is reflected.

wedge filter—bandpass filter with variable centre wavelength (between about 360 and 800 nm). Such filters are available in both linear and circular form. The actual wavelengths to be passed are based on the position of the wedge that is used.

multi-bandpass filter—several bands of the emission spectrum are transmitted. This is useful in fluorescence microscopy to simultaneously observe different dyes, e.g. DAPI/FITC or DAPI/FITC/TRITC.

3.2 Camera types and performance
3.2.1 Camera types

After having adjusted the microscope in such a way that high quality images are formed, it is important to choose an appropriate imaging device. In microscopy, CCD cameras and image-intensified systems are often used. Other image detectors are possible but the discussion of all possible camera types is beyond the scope of this chapter. Instead, several methods will be presented to evaluate the performance of a CCD camera which can be used for other detectors as well.

The vital part of the optical system of a camera consists of a CCD-array. This is the light-sensitive part of the camera and essentially consists of a thin silicon[1] layer. Photons reach the layer and break bonds between neighbouring silicon atoms. This generates electron-hole pairs. During the exposure time these (photo) electrons are collected in potential wells. A well can collect only a limited amount of electrons so the integration time has to be adjusted to prevent overflow. After integration the charges stored in the image section are read out and measured to determine the intensity of light. This is done by shifting each row sequentially into a serial register. Here one pixel after another is converted into a digital signal proportional to the charge. This signal can be transmitted and measured. It is important to protect the storage section from light during the readout process to prevent smearing. Several CCD architectures with different arrangements of image section and storage section have been developed: full frame transfer, interline transfer, and frame transfer. The full frame architecture is the only one that needs an external shutter to prevent further photon collection during readout. The other architectures can shift the charges to a light-protected region before shifting to the serial register (see *Figure 13*).

The analogue data received by the camera have to be transformed to digital data before they can be processed by a computer. An analogue-to-digital converter (ADC) has to be placed between camera and computer. Depending on the camera type different tools are possible. Some cameras have built-in interfaces to the computer while others need a frame grabber in between.

In the class of CCD cameras we distinguish between video and scientific cameras. Scientific cameras have the advantage that they are very sensitive and that they have optional settings that can be adjusted to the signal that has to be detected. The disadvantage is the usually high price compared to simpler video cameras. Video cameras can be sufficient for microscope imaging, depending on the specimen. They can be modified to allow variable integration time.

---

[1] Silicon is opaque to wavelengths smaller than 400 nm (UV light is blocked) and transparent to light of a wavelength above 1100 nm (IR light). Within this interval the penetration depth depends on the wavelength of the incident light. Green light penetrates about 1 μm, whereas red light with a longer wavelength penetrates 5–8 μm.

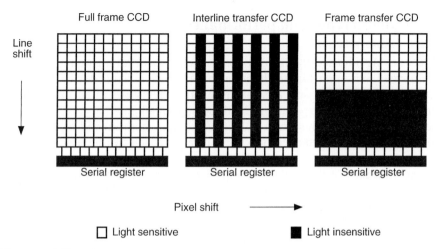

Figure 13. CCD architectures.

3.2.2 Characteristics

Many companies offer scientific CCD cameras or video-based CCD cameras that are suitable for microscope image acquisition. However, it is difficult to compare cameras of different manufacturers as every company gives a different set of specifications and often there is no explanation as to how the specifications have been measured. To determine the suitability of a camera type several aspects should be studied (18, 19):

Signal to-noise ratio—A number indicating the quality of the image is the signal-to-noise ratio (SNR). It can be calculated from two independent images I_1 and I_2 of a homogeneous, uniformly illuminated area with the same mean value μ. The variance σ^2 of the difference between the two images depends on photon shot noise, dark current, and readout noise (see Section 3.2.4) and is calculated as:

$$\sigma^2 = \frac{1}{2}\text{var}(I_1 - I_2) \qquad [4]$$

Using these values the SNR is calculated as:

$$\text{SNR} = 20 \log\left(\frac{\mu}{\sigma}\right) \text{ (dB)} \qquad [5]$$

The SNR should be at least 20 dB. Lower SNR means that the image is too noisy for measurements.

Ignoring possible noise sources with the exception of the inevitable Poisson noise mentioned in Section 3.2.4, it is possible to determine the best SNR a camera can achieve. Every CCD well has a finite capacity for photoelectrons, N. The maximal signal thus is N with a standard deviation of $\sqrt{N}$ according to Poisson statistics. The resulting SNR according to

Equation 5 is reduced to SNR = $10\log_{10}(N)$. Knowing the size of the pixels and the typical photoelectron density of modern cameras, which is about 700 e$^-$ μm^{-2}, the maximum SNR for a camera can be determined (19). For example, a camera with small pixels (6.8 × 6.8 μm^2) has a maximum SNR of $10\log_{10}$(6.8 × 6.8 μm^2 × 700 e$^-$ μm^{-2}) = 45 dB. Large pixels (23 × 23 μm^2) lead to a maximum SNR of 56 dB.

Pixel size—Small pixel sizes offer a higher sampling density than larger pixel sizes but the pixels are more limited with respect to noise. The SNR is related to the size of the pixels (18, 19) meaning that the larger the pixel the higher the possible SNR.

Fill factor—A number that indicates what percentage of the chip is actually light sensitive. For example, the interline transfer CCD chip and the frame transfer CCD chip can have a fill factor of at most 50% (see *Figure 13*).

Sensitivity—We distinguish between two types of sensitivity, the absolute and the relative sensitivity. The absolute sensitivity of a sensor is determined as the minimum number of photoelectrons that can be detected. This number depends on the noise characteristics of the camera. The relative sensitivity is the number of photoelectrons needed to produce a change of one analogue-to-digital converter unit (ADU) (18).

Spectral sensitivity—The silicon layer used to detect light is not equally sensitive to different wavelengths. Silicon is opaque to wavelengths smaller than 400 nm (UV light is blocked) and transparent to light of a wavelength above 1100 nm (IR light). Within this interval the sensitivity increases until it reaches its maximum at about 1050 nm as shown in *Figure 14*.

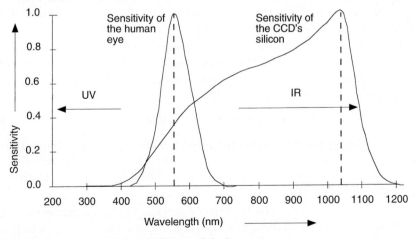

Figure 14. Spectral sensitivity of silicon and the human eye.

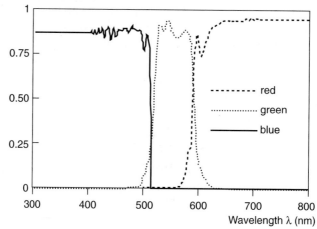

Figure 15. Spectral characteristics of the red, green, and blue filters used in RGB colour cameras. (Information supplied by Andover Corporation, Salem, NH, USA.)

For colour cameras the sensitivity is determined by the silicon sensitivity in combination with the filters used to separate the red, green, and blue signals. An example is shown in *Figure 15*.

Quantum efficiency—The relationship between the average number of incident photons and the average number of resulting photoelectrons is the quantum efficiency of the sensor material. Not every photon causes an electron hole pair. But as the arrival of photons is determined by Poisson statistics, the number of photoelectrons is also Poisson distributed. The quantum efficiency depends on the wavelength of the incident light. For silicon it is about 50%, less for wavelengths tending to the blue end of the spectrum, and more for wavelengths tending to the infrared end.

Linearity—Especially for intensity measurements it is desirable to get a linear response between the output signal and the input signal. The intensity of the output image depends on the intensity of the input image, the gain, the gamma value, and the offset:

$$I_{out} = gain \cdot I_{in}^{\gamma} + offset \qquad [6]$$

A linear response means that the offset is 0 and γ is 1. Unfortunately, the offset is almost never zero. There are several methods to correct for the non-zero offset after image acquisition. The most suitable method is discussed in Section 3.3.

The value of γ depends on the type of camera. Modern cameras offer the ability to switch electronically between various values of γ, which influences the contrast of the image. If γ is less than 1.0 the dynamic range

1: Microscope image acquisition

Figure 16. Test pattern with different realizations of γ (γ = 0.6, 1.0, 1.6 from left to right).

of grey values is compressed. For values of γ greater than 1.0 the dynamic range is expanded (see *Figure 16*). Some cameras have an automatic γ-correction to give the best possible contrast in the output image. This should be turned off if intensity measurements have to be done as different images will be acquired under different conditions and thus will not be comparable.

The gain indicates the amplification of the input signal. Cameras can have automatic gain control 132 (AGC) which sets the gain automatically to obtain best contrast in the image. This function should also be switched off to enable intensity comparison between different images. Sometimes it is possible to choose between different but fixed gain values, usually 1× and 4×, which can be used to adjust the camera sensitivity to the input signal. If the input signal is low the gain should be high.

Protocol 2. Calibration for intensity measurements (densitometry)

Digital images can be used to measure intensity information in the specimen. The grey value of the digital image can easily be related to properties such as optical density. If the system is not linear a calibration function has to be determined (17):

1. Take images of graded neutral-density filters (placed in the object plane of the microscope) and average the entire digital image.
2. Relate the known input intensity to the measured output intensity.

If the images are distorted by uneven illumination use flat field correction (see Section 3.3) to remove the shading.

Binning—A possibility to reduce noise is binning. This is a special read-out method available with some cameras. Instead of measuring every pixel it is possible to combine the charge of a number of neighbouring pixels before digitization. Thus the noise contribution is relatively reduced. The resulting images consist of a reduced number of pixels which makes processing faster but also decreases the spatial resolution as well as the sampling density.

3.2.3 Analogue signal to digital signal

The output signal of a camera can be either digital or analogue. Scientific CCD cameras usually give 8, 10, or 12 bit digital output. Video-based CCD cameras deliver an analogue signal that has to be transformed to a digital signal by a frame grabber. Depending on the specifications of the frame grabber the resulting image differs in number of pixels in the horizontal direction from the expected number based on the camera chip specifications. In the following section the sampling of an image by the camera is discussed. If a frame grabber is involved in the image acquisition the combination of camera and frame grabber determines the sampling, not the camera alone.

When we want to measure quantities in images recorded by a camera it is important to know how the original image is sampled by the camera pixels. The numbers of pixels in the horizontal and vertical directions indicate the resolution of the camera. These numbers are given by the manufacturer. The distance between two pixels on the CCD-array corresponds to a certain distance in the imaged object. It is possible that the sampling density (SD) in the x-direction of the chip differs from the SD in the y-direction. To determine the SD of the optical system a test pattern with known dimensions is needed. The actual size of the test pattern is compared to the size in pixels in the image.

$$SD = \frac{\text{size \{(in pixels)\}}}{\{\text{size (in } \mu m)\}} \qquad [7]$$

This has to be done in the x- and y-direction of the camera chip. The pixels of a camera are square, only if

$$\frac{SD_x}{SD_y} = 1 \qquad [8]$$

It is not possible to produce an image that is identical to the original object. Due to the finite opening of the objective only part of the light scattered or emitted from the object is collected. Part of the information about the specimen is lost with the lost light. Considering the effects in the Fourier domain the lens can be characterized as a bandpass filter where only frequencies between $-2NA/\lambda$ and $2NA/\lambda$ are transmitted; other frequencies are suppressed. Therefore, the frequency $2NA/\lambda$ is called the cut-off frequency. An image that is to be processed in a computer has to be sampled at or above the *Nyquist frequency*, that is twice the cut-off frequency:

$$SD_{\text{Nyquist}} = \left\{\frac{4NA}{\lambda}\right\} \left\{\frac{\text{pixels}}{\mu m}\right\} \qquad [9]$$

1: Microscope image acquisition

Only then it is possible to reconstruct the original image based on the sampled image and avoid aliasing. Depending on subsequent image-processing steps, higher sampling may be necessary (20).

Assuming sampling at the Nyquist frequency we can determine the required SD for the two model objectives 25×/0.6 and 40×/1.3 Oil assuming a wavelength of 500 nm. The minimum required sampling density is $4NA/\lambda$ resulting in SD = 4.8 pixels per μm for the first objective and 10.4 pixels per μm for the second.

If the aim of image acquisition is to analyse the image content it is advisable to sample at a higher frequency than proposed by the Nyquist theorem. This theorem is appropriate to yield images that *look* good. For image analysis these images, however, may be unsuitable. The accuracy of measurements in the image highly depends on the sampling density (21).

Protocol 3. Sampling rules of thumb

For measurements in images with an allowable error of about 1% the following numbers can be used as a rule of thumb (21):

- 10–20 samples per average cell diameter to measure the area.
- 25 samples per analogue length to measure length.
- 150 samples per average cell diameter to measure certain texture features.
- 15 to 30 samples per μm (2–4 times Nyquist frequency) to measure small structures of the order of micrometres.

The sampling density highly depends on the maximum allowable error of the measurement and the measurement method.

3.2.4 Noise sources

The image obtained by a camera may suffer from several unwanted contributions. These can be stochastic (random noise) and/or systematic. It is desirable to suppress the distortions as much as possible. While it is quite easy to correct for additive systematic distortions, it is very difficult to remove random noise from an image. Steps that reduce or prevent noise in advance, such as cooling a camera to suppress thermal agitation of electrons, are preferable.

Several noise sources may be contained in the camera itself, such as:

Photon noise—caused by quantum nature of light. Arrival of light particles—photons— at a detector is a stochastic process with a Poisson distribution. This kind of noise cannot be suppressed. The best possible image still shows photon noise.

Dark current—the term for the noise in a camera image that is caused by thermal agitation of electrons instead of light excitation. If the shutter is closed during integration the resulting image is the dark current. Dark current increases with the temperature of the camera. Scientific cameras are provided with a cooling unit which keeps the operating temperature of the camera low to suppress dark current.

Readout noise—noise added during readout of charges by the camera chip electronics. The amount of noise depends on the readout rate. For scientific cameras with a moderate readout rate (20–500 kHz) the readout noise is minimal. Images obtained with video-based cameras with high readout rates show significant noise contributions due to the on-chip electronics.

Quantization noise—noise generated during conversion of charges to digital data. The ADC converts analogue data to integer numbers. The round-off error occurring during this conversion is responsible for the quantization noise. The amount of noise depends on the number of bits. A standard formula to determine the SNR of quantization noise is

$$\text{SNR} = 6b + 11 \text{ (dB)} \qquad [10]$$

with b the number of bits used for the digital representation of the data. If $b = 8$ the amount of quantization noise is negligible.

Blooming—appears if a potential well is saturated with charge. Furthermore, electrons can be passed to neighbouring wells which distort the result. The exposure time should be adjusted. No pixel in the output image should have a maximum grey value as it is impossible to judge if neighbouring pixels might have collected a surplus of photoelectrons of the saturated pixel (see *Figure 17*).

Smearing—effect occurs if light reaches the CCD-array during readout.

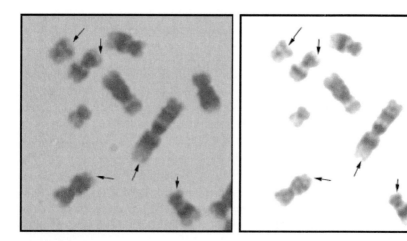

Figure 17. Metaphase chromosomes. *Left*, correct exposure time; *right*, overexposed. The arrows point at some of the regions where information is lost if the exposure time is too long.

1: Microscope image acquisition

The CCD array has to be protected during readout. A shutter can be inserted in the optical path to block the light path as necessary.

Dead pixels—pixels that do not react to incoming light but which give a constant white or black value. The manufacturer determines the quality of each CCD chip after production. The fewer dead pixels a chip has the better its quality.

Protocol 4. Considerations in choosing a camera

To find the appropriate camera for a specific application the following should be considered:

1. What is imaged? In particular, what is the smallest detail that has to be detected, what is the largest object, what is the wavelength of light, what is the amount of light, how sensitive is the specimen to excitation, how many objects have to be imaged, and how much time may the imaging procedure take?

2. How is the image obtained? What kinds of objective, filters, relay lens, etc., are present in the optical path, and how do they change the image-forming light, especially the intensity and the wavelength?

3. What is the purpose of the image? Is the image used for measurements or for visualization?

The answers to these questions help to specify the requirements of the camera. Note that the image quality is mainly determined by the *optics* between specimen and camera and not by camera specifications.

To illustrate the considerations to determine the requirements of a camera let us assume we want to image fluorescent cells with a diameter of 20 μm stained with a red dye ($\lambda = 620$ nm). The cells include a cell nucleus with a diameter of 5 μm stained with a blue dye ($\lambda = 460$ nm). We are interested in the area of the cell and the nucleus.

The image is mainly controlled by the objective lens of the microscope. The magnification determines the size of the image that is presented to the camera and the numerical aperture determines the amount of light that is present. It is advisable to use a magnification as low as possible and a numerical aperture as high as possible. Then the field of view is large and resolution and light level are high. In fluorescence microscopy, where the signal is weak, high NA objectives are essentially for good images. In this case a 40×/1.3 Oil immersion lens would be recommended.

To detect the signal of interest we insert filters in the optical path: one to transmit the red colour, the other to transmit the blue colour. A dual bandpass filter enables the imaging of both signals in one image. However, since the CCD silicon is much more sensitive to the red signal than to the blue signal (see *Figure 14*), the red signal should be slightly suppressed to prevent saturation of camera pixels before sufficient blue signal is detected.

The sampling is determined by the smallest object to be inspected and by the measurements that have to be performed. In this case the cell nucleus with a diameter of 5 μm should be represented by 20 pixels (see *Protocol 3*). With this knowledge we can compute the maximum pixel size of the camera chip. The image that is presented to the camera is magnified by 40. The diameter of the cell nucleus is then 40 × 5 = 200 μm. To represent this by 20 pixels the size of the camera pixels must be 200 μm /20 pixels = 10 μm per pixel. Thus the actual sampling density is 20 pixels/5 μm = 4 pixels per μm if this size chip exists. Comparing this to the theoretical Nyquist rate of 4 × 1.3 pixels/ 0.46 μm = 11.3 pixels per μm we see that we severely undersample. However, the accuracy of the area measurement is less sensitive to Nyquist sampling [20]. Therefore, in this case it is not necessary to use a camera with smaller pixels. If sampling at Nyquist is required, the camera pixel size would have to be 2.7 × 2.7 μm^2. This is a very small pixel size that will rarely be found in CCD cameras. To reach Nyquist sampling it is also possible to increase the magnification by inserting a zoom lens in the optical path.

Considering the maximum SNR of the chosen camera we get SNR = 10 × $\log_{10}$(10 μm × 10 μm × 700 e$^-$/μm^2) = 48 dB which is sufficient for measurements.

The camera should have the option of integrating. Fluorescent signals are usually of low intensity. To obtain a good image we have to accumulate photons in time. The longer the integration time the more important it is to use a cooled CCD camera with a slow readout rate.

We now have the 'minimum' constraints for our camera depending on the smallest detail we want to inspect. However, there is also a 'maximum' constraint. For measurements on the cell with a diameter of 20 μm it is necessary that the whole cell fits within the field of view. The cell diameter will be imaged on 20 μm × 4 pixels/μm = 80 pixels. To image one cell completely the camera chip should have at least 100 × 100 pixels. This is not a problem in practice.

3.3 Images

The image quality does not depend only on image acquisition. Careful specimen preparation can yield better images than any image-processing technique applied to images of a poor preparation. For example, for the spot counter it is desirable to have many non-touching cells in one image to get information on a large number of cells in a short time (3). Another important point is to have clean specimen preparations. Even though the human eye succeeds in separating desired objects from debris it can be very difficult to do this automatically. Any source of debris, such as dust particles, should be avoided. The preparation technique should also ensure reproducibility. Automated systems are designed for a specific class of input signal. The performance characteristics

1: Microscope image acquisition

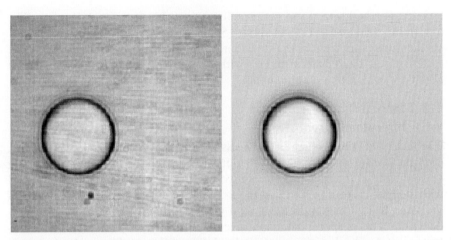

Figure 18. Original (*left*) and flatfield (*right*) corrected image of a microbead.

of the system can only be guaranteed if the input signal does not vary too much. Care must be taken in specimen preparation.[2]

Even with clean specimen preparation and optimal optics it is possible to get undesired contributions in the output image, such as shading or dark current. Shading is due to an uneven illumination of the field of view. To eliminate deterministic variations in an image, every image taken by a camera should be corrected by flat field correction. Next to the image of the object, an image of the background is needed (blank image) and an image of the camera 'noise' (dark current). The γ of the camera and the exposure times for all images should be the same. The corrected image is (see also *Figure 18*):

$$I_{\text{Corrected}} = \frac{I_{\text{object}} - I_{\text{dark}}}{I_{\text{blank}} - I_{\text{dark}}} \qquad [11]$$

Note that the image brightness produced by this correction will be floating point numbers between zero and one.

Further improvement of the image can be achieved by applying special image-restoration techniques (22). The basic idea behind these techniques is to invert the imaging process to obtain the original, noise-free object. (See also Chapters 2 and 3 on image enhancement.)

4. Application

4.1 Objective

The spot counter project was set up to develop a system that could count chromosome spots in interphase nuclei, either in brightfield or in fluor-

---

[2] As Professor T. S. Huang of the University of Illinois once said, 'The best image restoration is a priori image restoration.'

escence, at a high speed and with results comparable to manual counting. The aim was to analyse 500 cells in about 15 min (2–4).

In the following section the considerations involved in the design of the system in refs 2 and 3 are explained with emphasis on the information we have about optics, cameras, and imaging.

4.2 Materials and methods

4.2.1 Instrumentation

The heart of the system is a Zeiss Axioskop microscope (Carl Zeiss, Oberkochen, Germany). It is fully automated to allow automated scanning and focusing of the specimen. This involves a fully automated stage equipped with stepper motors to move in the x-, y- and z-direction. The resolution of the stage is 0.1 μm/step in the x- and y-direction and 0.04 μm/step in the z-direction. The proper excitation wavelength can be adjusted with an automated filter wheel with space to hold five different filters. A shutter is installed in front of the 100-W mercury lamp to interrupt the excitation when possible to prevent bleaching of the fluorescent specimen. These instruments are provided by Ludl Electronic Products, Ltd, Hawthorne, NY, USA (2, 3).

The images are acquired with a KAF 1400 Photometrics Series 200 camera (Photometrics Ltd, Tucson, AZ, USA). This is a slow scan (500 kHz), cooled (–42°C) CCD camera. The chip consists of 1317 × 1035 pixels. Each pixel has a size of 6.8 μm × 6.8 μm. On-chip binning is possible. Other noise sources besides photon noise are insignificant with this camera (18). In fluorescence the signal is small, so it is important to use a sensitive camera with excellent noise suppression.

All instruments are controlled by a Macintosh Quadra 840 AV computer (Apple Computer Inc., Cupertino, CA, USA). The connection of camera and computer is by a NuBus interface. The other components are connected via one serial line of the Macintosh computer.

The choice of a suitable objective depends on a number of constraints. As the processing of the specimen has to be fast, it is advantageous to use a low magnification as the field of view is larger than with a higher magnification. For the two model objectives 25×/0.6 and 40×/1.3 and assuming sampling at Nyquist frequency, the 25×/0.6 lens covers a field of view of 274 μm × 216 μm. That is more than four times larger than the field of view seen with the 40×/1.3 lens (127 μm × 100 μm).

Another point of consideration is image brightness. The image brightness is (in theory) proportional to the NA^4 (17).[3] In the case of the model objectives this means that images acquired with the 40×/1.3 Oil objective are more than 20 times brighter than images acquired with the 25×/0.6 objective given equal integration time. Or, to get the same image intensity, the integration time with

---

[3] Piper (personal communication, 1997) has reported that in practice the image brightness in an epi-illumination fluorescence microscope is proportional to the NA^2. In any event image brightness is strongly affected by the NA.

the 25×/0.6 objective has to be 20 times the integration time necessary for the 40×/1.3 Oil lens.

Furthermore, the systems require high resolution as the dots are very small (diameter ~ 1–2 μm). In Section 2.3 we saw that the resolution for the model objectives is 0.51 μm and 0.23 μm respectively. The 40×/1.3 Oil objective thus is better with respect to resolution than the 25×/0.6 objective.

Another problem is that the DOF decreases with increasing NA. As the cell nuclei are three-dimensional objects the spots can lie in many different planes. If we focus on one spot, other spots may be out of focus or even invisible if they are far from the focal plane. The DOF at $\lambda = 500$ nm is 0.69 μm and 0.22 μm respectively for the above-mentioned objectives, 25×/0.6 and 40×/1.3 Oil.

On balance, the 40×/1.3 Oil objective is more suited for the spot counter system. The time to acquire sufficient signal is much smaller than with the 25×/0.6 objective. The time necessary to scan the complete slide is still less compared to the 25×/0.6 objective even though more fields of view have to be scanned. The resolution is also better which makes it easier to detect spots and perform measurements on these small objects.

The second choice that has to be made is which filter should be used. If only two colours are involved, one to mark the chromosomal part of interest and the other to counterstain the DNA, a dual bandpass filter or two single filters can be used. The dual bandpass filter maps the background to black, the counterstained nuclei to medium grey, and the spots to white. The dynamic range of the camera is sufficient to be able to distinguish between the three possible classes. The use of single bandpass filters would have the advantage that the contrast is larger as only background and one specific colour are visible in the image. But this would also mean that two images instead of only one image would have to be acquired, stored, and analysed. The choice of the filters thus depends on the number of colours that should be detected. In this case, where only two colours are involved, the dual bandpass filter was used (see *Figure 19*).

4.2.2 Scanning cycle

The completely automated microscope system to count (fluorescent) dots repeats a number of steps until a predetermined number of nuclei has been analysed or the slide is completely scanned. During this scanning cycle the stage is moved to the new position, the field of view is focused automatically, the image is acquired and analysed, and the result is updated.

Autofocus—As the slide has to be scanned automatically a procedure is needed to find the correct in-focus position of the stage automatically. Boddeke *et al.* (23) describe a procedure that makes use of a special focus function. This focus function $F(z)$ has the following properties:

- The maximum of the function is reached at the in-focus position.
- Around the maximum is a sharp peak.

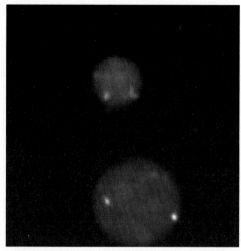

Figure 19. Image of two interphase cell nuclei with spots obtained with a dual bandpass filter.

- The function has only one maximum.
- It is easy and fast to calculate.
- It has a wide focus range.

One possibility to obtain such a function $F(z)$ is the following. After applying a $\{-1,0,1\}$ digital filter in the x-direction the 'energy' in the image $I(x,y,z)$ can be computed (2):

$$F(z) = \sum_{x,y}(I(x+1, y,z) - I(x-1, y,z))^2 \qquad [12]$$

The determination of an optimal in-focus position of a stage is done in three steps. The first step is to find the so-called sloped region, the region where the focus value has a maximum. With the second step this region is further enclosed. In the third step N samples are taken within the region of the maximum. A quadratic function is fit through these samples. The maximum of the function is used as an estimate of the focus position and the stage is moved accordingly.

Depending on the objects that are inspected it is advantageous to use binning before calculating the focus function $F(z)$ to increase the processing speed and the SNR. The optimal binning factor has to be determined empirically. In the case of the spot counter a binning factor of 12×12 was found to be preferable.

Protocol 5. Focus function for autofocus

A simple way to implement a focus function is the following:

1. Acquire an image $I(x,y,z)$.

1: Microscope image acquisition

2. Generate two new images by shifting $I(x,y,z)$ one pixel to the left ($= I(x-1,y,z)$) and one pixel to the right ($= I(x+1,y,z)$).
3. Subtract the two new images from each other.
4. Square the result and add all pixel values.
5. Compare this result with results obtained at other focus positions. Maximize $F(z)$ over z to focus the image.

Image analysis—after acquiring the image it has to be analysed by the computer. The image analysis procedure of the spot counter is divided into four major steps:

- Find the regions of interest (ROIs)—only about 5% of the image contains relevant information. Everything else is background. Instead of searching for nuclei in the whole image the search is performed on a smaller image. By combining a number of neighbouring pixels of the original image into

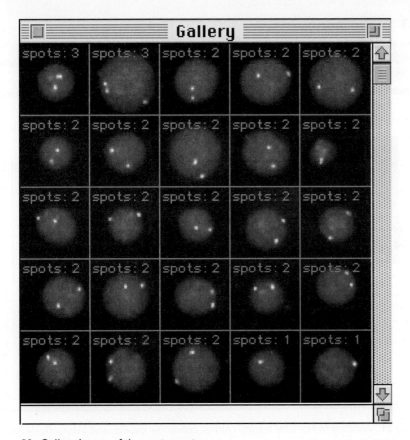

Figure 20. Gallery image of the spot counter.

one pixel of a reduced image, fine details are lost but large objects, such as the cell nuclei, are still visible. This image is sufficient to find the ROIs. As the number of pixels is decreased the search is much faster.
- Find the nuclei within the ROIs—in the previously determined ROIs the cell nucleus is found. This is done at the original resolution, not in the reduced image.
- Find spots in cells—only spots that lie in cells have to be counted. In the previously determined cell nucleus the spots are found.
- Count spots and update results—all cell nuclei found and their containing spots are displayed in a gallery and the counting results are stored in a file.

After the analysis has finished the user can scroll through a gallery of acquired cell images (*Figure 20*) and check the results. In case of doubt he or she can choose the particular cell image. The stage automatically moves to the position where that cell has been found. The user can observe the cell through the microscope or on the computer monitor and correct the result as necessary.

4.3 Performance

The spot counter described is able to analyse 500 interphase cell nuclei in about 15 min. The result of the spot counter analysis is comparable to manual

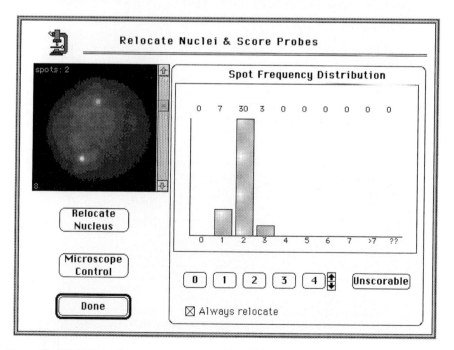

Figure 21. Window to relocate nuclei, control computer results, and interactively correct the individual dot count.

counting. About 89% of the cell nuclei are counted correctly by the system. Only 11% of the results have to be corrected by the operator.

The scanning time depends on the number of cell nuclei per field of view, on the NA of the objective used, on computer speed, and on the camera readout rate. If any of these conditions is changed the scanning time will change as well.

The user-friendly interface allows interaction at the end of the scanning. In addition, the main results of the analysis are displayed. The user can correct the computer result as necessary (*Figure 21*). The spot counter result is presented in a histogram. This histogram shows the computer results versus the interactively corrected results. The user can see how the system performed and what changes were done during correction in a confusion matrix (see Chapter 4).

References

1. Swift, H. (1966). In *Introduction to Quantitative Cytochemistry* (ed. G. L. Wied), pp. 1–39. Academic Press, New York.
2. Netten, H., Young, I. T., van Vliet, L. J., Tanke, H. J., Vrolijk, H., and Sloos, W. C. R. (1997). *Cytometry*, **28**, 1–10.
3. Netten, H., Vrolijk, H., Sloos, W. C. R., Tanke, H. J., and Young, I. T. (1996). *Bioimaging*, **4**, 93–106.
4. Vrolijk, H., Sloos, W. C. R., van der Rijke, F. M., Mesker, W. E., Netten, H., Young, I. T., Raap, A. K., and Tanke, H. J. (1996). *Cytometry*, **24**, 158–166.
5. Castleman, K. R. (1996). *Digital image processing*, 2nd edition. Prentice-Hall. Englewood Cliffs.
6. Jähne, B. (1997). *Practical handbook on image processing for scientific applications*. CRC Press, Boca Raton, Florida.
7. Bradbury, S. (1984). *An introduction to the optical microscope* (Microscopy Handbooks 01). Royal Microscopical Society, Oxford University Press.
8. Batchelor, B. G., Hill, D. A., and Hodgson, D. C. (1985). *Automated visual inspection*. IFS (Publications) Ltd, UK.
9. Pawley, J. B. (1995). *Handbook of biological confocal microscopy*, 2nd edition. Plenum Press, New York.
10. Wilson, T. (1990). *Confocal microscopy*. Academic Press, London.
11. Goodman, J. W. (1996). *Introduction to Fourier optics*, 2nd edition. McGraw-Hill, San Francisco, USA.
12. Rost, F. W. D. (1992). *Fluorescence microscopy*, Vol. 1. Cambridge University Press.
13. Rost, F. W. D. (1992). *Fluorescence microscopy*, Vol. 2. Cambridge University Press.
14. Nikon technical bulletin (1992). *How to use a microscope and take a photomicrograph*. Nikon Corporation.
15. Leitz GmbH (1997/Jena). *Image-forming and illuminating systems of the microscope*. Ernst Leitz GmbH, Wetzlar, Germany.
16. Kapitza, H. G. (1997). *Microscopy from the very beginning*. Zeiss brochure.

17. Inoue, S. (1986). *Video microscopy*. Plenum Press, New York.
18. Mullikin, J. C., van Vliet, L. J., Netten, H., Boddeke, F. R., van der Feltz, G. W., and Young, I. T. (1994). *Methods for CCD camera characterization* (Proc. SPIE Conference, San Jose, CA, USA, 9–10 February 1994). SPIE, **2173**, 73–84.
19. van Vliet, L. J., Boddeke, F. R., Sudar, D., and Young, I. T. (1998). In *Digital image analysis of microbes* (ed. M. H. F. Wilkinson and F. Schut). Wiley, pp. 37–64, UK.
20. van Vliet, L. J. (1993). *Grey-scale measurements in multidimensional digitized images*. PhD Thesis, Delft University Press, The Netherlands.
21. Young, I. T. (1988). *Analytical and Quantitative Cytology and Histology*, **10**(4), pp. 269–275.
22. van Kempen, G. M. P., van Vliet, L. J., Verveer, P., and van der Voort, H. T. M. (1997). *Journal of Microscopy*, **185**(3), pp. 354–365.
23. Boddeke, F. R., van Vliet, L. J., Netten, H., and Young, I. T. (1994). *Bioimaging*, **2**, 193–203.
24. Determann, H. and Lepusch, F. (19xx). *The microscope and its application*. Ernst Leitz GmbH, Wetzlar, Germany.
25. Young, I. T. (1989). In *Methods in cell biology* (ed. D. L. Taylor), pp. 1–45. Academic Press.

2

Biological image processing and enhancement

G. W. HORGAN, C. A. REID, and C. A. GLASBEY

1. Introduction

This chapter considers simple ways to enhance digital images, both to make visual interpretation easier and to make the images more amenable to subsequent automatic analysis. It is typically the stage in image analysis that follows image capture. Methods include adjustment of brightness or contrast, edge detection and edge sharpening, removal of small irrelevant features (termed noise), and adjustment for background intensity variation.

Image enhancement can only improve and clarify information that is already present in an image: it cannot recover what was never there in the first place! Therefore, at the stage of image capture it is essential to seek the best image quality by suitable sample preparation, adjustment of instrumentation, choice of lighting, etc. We state this as a principle:

> It is better, and often easier, to improve quality at the stage of image capture rather than by subsequent processing.

Image enhancement transforms a given image to a new image, where pixel intensities in the new image are functions of the pixel intensities in the original image. In this chapter we describe four classes of processing operations. In the simplest, which we will term *contrast manipulation* (Section 2), each new pixel intensity is a function of the old intensity at that pixel alone. We then describe *filtering* operations (Section 3) where the new intensity is a function of the old intensity at neighbouring pixels also. In Section 4, we consider a distinctive class of image-filtering techniques, termed *mathematical morphology*. We conclude, in Section 5, by describing operations for identifying local *textures* in images.

We will use three images in this chapter to illustrate the various operations we describe. The image in *Figure 1a* shows fungal mycelia, grown as part of a study of factors affecting their development (1). Pixel intensities range from 0

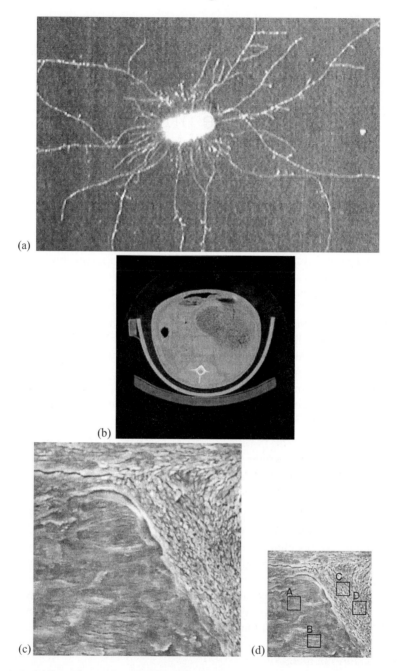

Figure 1. (a) Fungal mycelia. (b) X-ray CT scan of a sheep. (c) Microscope image of bone. (d) The bone image with four subimages indicated.

to 255. The large object near the centre is a food source originally containing the fungus, and the mycelia are the threadlike objects searching for other food sources. *Figure 1b* shows an X-ray CT image of a sheep, lying upside down in a crate. Various internal organs may be seen. Pixel intensities range from 0 to 2048. *Figure 1c* will be used for discussing texture. It shows a rat's tibia, viewed using a scanning electron microscope. There are several texture types that can be recognized on the surface of a bone such as this, each representing a different stage in the bone's mineralization process. The example shown consists of two types. Pixel intensities range from 0 to 255.

It is essential before enhancing an image to have a good idea of its original quality, i.e. level of noise, sharpness, background variation, etc. In *Protocol 1* we recommend steps to be followed in an initial image inspection.

Protocol 1. Inspecting an image

1. Magnify areas of the image where pixel values appear to be constant (see zoom in Section 2), to evaluate how much noise is present. Examine the range of pixel intensities.

2. Magnify areas of the image where there are boundaries between regions of constant pixel value. How distinct are the boundaries? The positioning of boundaries is important if the areas of regions are to be measured.

3. Look for variations in intensity across the image. These may cause problems in processing and analysis. (Removal of background variation is considered in Section 4.)

4. Consider edge effects in the image. Parts of objects that overlap the edge of the image will be missing, and measurements of properties will be affected. Some strategy for dealing with this will need to be devised and this will depend on the nature of the image analysis proposed.

---

2. Contrast manipulation

The equipment used for image capture will deliver an image with a range of pixel values. It may be helpful for interpretation or processing to first transform these intensities. This may have the effect of expanding the whole intensity range (the contrast) of the image for easier examination, or it may emphasize intensity variation in some important part of this range. Such transformations are based on individual pixel values only, and their spatial arrangement is not involved. The transformed pixel intensity is simply a function of the original intensity.

We classify the possibilities for intensity transform into four types.

- In *functional* intensity transform, the intensities are transformed by some user-specified function. This is analogous to the transforms, such as the logarithm, that may be applied to other scientific data.
- With *histogram-based* transforms, the original distribution of intensities is transformed to make best use of the range available.
- A *pseudocolour* display maps the intensities into colour space.
- *Thresholding* transforms produce a binary image, a much-simplified version of the original.

A contrast manipulation may transform the stored pixel intensities, or it may just control the image display. The way it is done will be determined by the image-processing software.

2.1 Functional transformation

The transformed intensity is defined to be some function of the original intensity. We may represent this as a plot of the function. *Figure 2* shows some examples. Image-processing software often provides similar plots. *Figure 2a* is the simplest transform—a linear function. The intensities between two limits are expanded to fill the full range. The limits may be chosen by trial and error or by examining pixel intensities of features in the image which are desired to be at the extremes of intensity in the transformed image. *Figure 2b* is a piecewise linear function—it consists of connected linear segments. Software often allows the user to move the junctions manually. *Figures 2c* and *2d* show logarithmic and exponential functions, which selectively enhance lower and higher intensities, respectively.

Figure 3 shows the X-ray CT image after a piecewise linear transform, which stretched all intensities from 768 to 1280 to occupy the full intensity display range, 0 to 255. Many of the structures can be seen much more clearly. This version of the image will be used in most subsequent processing.

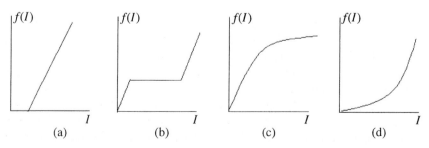

Figure 2. Intensity transformation: (a) linear; (b) piecewise linear; (c) logarithmic; (d) exponential.

2: Biological image processing and enhancement

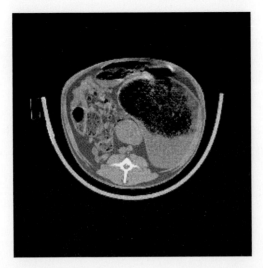

Figure 3. X-ray CT image after piecewise linear intensity transformation.

2.2 Histogram-based transformations

Histogram-based transformations use the histogram of pixel values to derive transformations which are in some way optimal. The most common is histogram equalization, a transformation which ensures that all display intensities are approximately equally represented. The intention is that ranges of pixel values are allocated portions of the display intensity range according to the frequency with which they occur in the image. This makes it easier to see intensity variations in the parts of the image corresponding to the histogram peak or peaks.

Let $p(I)$ denote the proportion of the pixel values $\leq I$. Pixel values of I are transformed to intensity $f(I)$, where

$$f(I) = 255 p(I)$$

The proportion of pixels with display intensity $\leq I$ is $I/255$, leading to a linear cumulative distribution of intensities, i.e. a uniform distribution. In practice, the discrete (i.e. non-continuous) nature of the image histogram will mean that the transformed image will only have an approximate uniform distribution. For a more detailed discussion of histogram equalization, see ref. 2.

Figure 4 shows a histogram-equalized version of the non-black regions in *Figure 3*. The contrast between some tissue types is more apparent here than in *Figure 3*. It was found more useful to equalize *Figure 3* than the original, which contained unimportant variation in the background. *Figure 5* shows corresponding histograms.

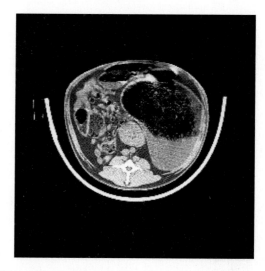

Figure 4. X-ray CT image (as in *Figure 3*) after histogram equalization.

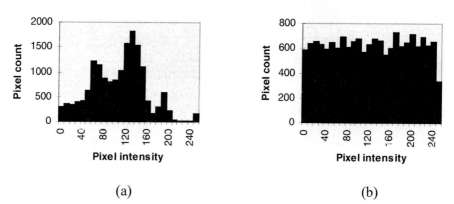

Figure 5. Histograms of (a) *Figure 3* and (b) *Figure 4*. The large number of black zero intensity pixels have been omitted.

2.3 Pseudocolour

The human eye finds it easier to compare colours than variations in intensity. With colour, our perception is less sensitive to context than it is with grey-levels and the illusion whereby an intensity is perceived relative to nearby intensities (e.g. a grey square surrounded by white appears darker than the same grey square surrounded by black) is less of a problem. Colour, which is essentially three-dimensional, also provides a richer space in which to present variation. The three-dimensional nature of colour is a biological artefact: our eyes have receptors sensitive to light in the red, green, and blue parts of the spectrum, and all colours are perceived in terms of the response they produce

2: Biological image processing and enhancement

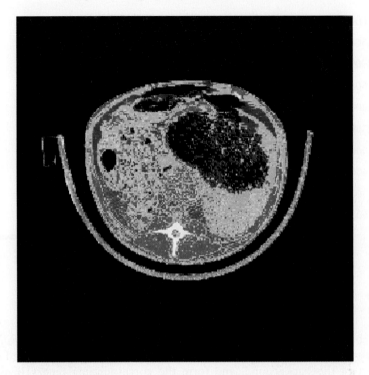

Figure 6. Pseudocolour display of the histogram-equalized CT image.

in these receptors. Most technological use of colour (computer screens, photographic processing, etc.) is based on this red, green, blue (RGB) system. A colour as displayed on a computer monitor may be represented as a position within the colour cube, a three-dimensional graph of the RGB intensities which can each vary between zero and some maximum.

A pseudocolour display of intensities takes a greylevel, monochrome image and maps the intensity range to some path in the colour cube. This is illustrated for the histogram-equalized CT image (*Figure 4*) in *Figure 6*. The lowest intensities are black, and the colour path then visits the blue, green, and red vertices of the colour cube before finally showing the brightest intensities as white. This is just one of many possible paths.

2.4 Thresholding

Many images have bimodal histograms of pixel values, because pixels essentially take only two different values, typically corresponding to objects and background. The fungal image (*Figure 1a*) illustrates such a situation. Often in image analysis we are interested in the properties of the objects: their geometry, shape, size, etc., and not in the variations in intensity within the objects

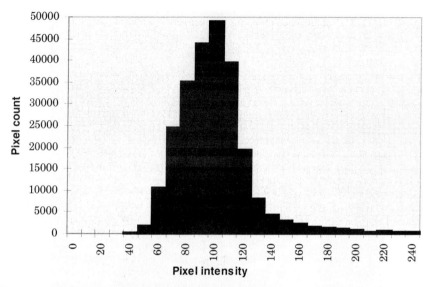

Figure 7. Intensity histogram of fungal image.

or the background. The most natural way to deal with this situation is to decide which pixels are part of one image component (the objects) and which are part of the background, and form the corresponding binary image for further analysis.

The simplest way to form a binary image is by 'thresholding': pixels with intensities less than or equal to the threshold become one component of the binary image, and those with intensities greater become the other. Sometimes, to allow more flexibility, we may assign to one of the components all pixels between two intensity limits (a double threshold).

The question immediately arises as to how the threshold(s) should be chosen. It is usually helpful to examine a histogram of pixel intensities. Ideally this will show two peaks, one corresponding to pixel intensities in the objects, the other to the background. A threshold may then be chosen to best separate the peaks. *Figure 7* shows a histogram of the fungal image in *Figure 1a*. Here, there is a single peak, and no clear separation between fungal and background pixel intensities. The peak is asymmetrical, being skewed towards higher values, and some of this skewness will be due to the mycelial pixels. Choosing a threshold here is more difficult.

Figure 8 shows the result of trying a range of thresholds. None is completely satisfactory, and it is not clear which is best. A decision on which to use will depend on what further processing is to be carried out. *Protocol 2* offers some suggestions for threshold choice.

Note that it is may be best to process the image, for example by background subtraction or filtering before thresholding.

2: Biological image processing and enhancement

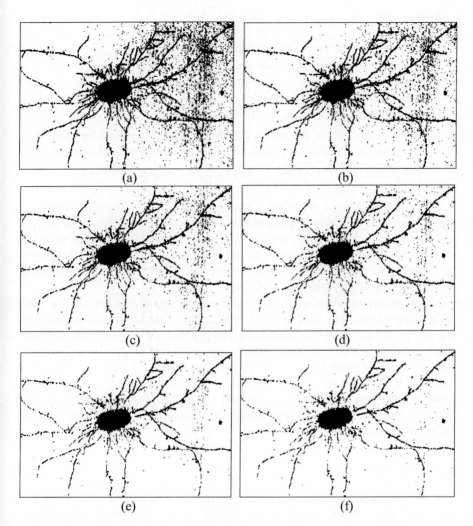

Figure 8. Binary fungal image derived using intensity thresholds of (a) 120, (b) 125, (c) 130, (d) 135, (e) 140, (f) 145.

2.5 Zooming

When an image has fine detail, we may not be able to examine this detail easily in a standard display on a monitor. We would like to enlarge a part of the image. We cannot usually enlarge the pixels on the monitor, and zooming must consist of allocating more than one pixel on the monitor for each pixel in the image.

> **Protocol 2.** Choosing a threshold
>
> We suggest the following approach to choosing a threshold.
> 1. Construct and examine a histogram of pixel intensities.
> 2. If the object pixel intensities clearly form a separate group a threshold may be selected, either manually or automatically. A review of automatic approaches may be found in ref. 3.
> 3. If there is no clear grouping, the threshold will need to be chosen manually. It may be more useful to examine the effects of a range of thresholds directly on the image.

The simplest form of zooming is pixel replication. For example, if we display an image pixel using a block of 2 × 2 pixels on the display, then we will apparently have doubled the size of the image. Obviously, no more detail is present, although it may be easier to see what is there. There may no longer be room on the display for the whole image. Zooming is equivalent to holding a magnifying glass in front of the screen.

Interpolation provides a method which produces smoother results with fewer distortions than simple pixel replication. The basic idea is that we regard the digital image as a discrete version of some continuous variable, which we have sampled at points on a lattice. Pixel locations in the new image will fall at different positions on the continuous surface. We do not, of course, know what the pixel values at these positions should be, but we may estimate them by assuming that there is a smooth change in pixel value between the four nearest pixels in the original image, and use bilinear interpolation. (See, for example, ref. 4, Section 2.4.)

3. Filtering

The operations described in the previous section modified the intensity at a pixel independently of other pixels. Other types of enhancement can be obtained by replacing a pixel with some function of the pixel intensities in the neighbourhood of that pixel. These operations are termed filters. We classify these into three types. Linear filters are the simplest and most widely available. They enjoy the advantage of having well-understood properties. Non-linear filters are more flexible and powerful, but need to be used with more care. Morphological filters derive from an elegant collection of ideas based on set theory and will be discussed in Section 4.

3.1 Linear filters

In a linear filter, a pixel is replaced with a linear combination of intensities of neighbouring pixels. The simplest neighbourhood is the 3 by 3 grid centred on

2: Biological image processing and enhancement

the pixel. We may represent the coefficients of the linear combination as a matrix:

$$\begin{matrix} a & b & c \\ d & e & f \\ g & h & i \end{matrix}$$

If pixel positions in an image are labelled (i,j), where i and j are row and column indices respectively, and the pixel intensity at (i,j) is denoted I_{ij}, then the filter output at (i,j) will be

$$\begin{aligned} &+ aI_{i-1,j-1} + bI_{i-1,j} + cI_{i-1,j+1} \\ &+ dI_{i,j-1} + eI_{i,j} + fI_{i,j+1} \\ &+ gI_{i+1,j-1} + hI_{i+1,j} + iI_{i+1,j+1} \end{aligned}$$

Smoothing filters

The simplest linear filter has all the weights the same, equal to one-ninth for the 3 × 3 neighbourhood. This smoothes the image, or can be considered as blurring it. It is a simple moving average. *Figure 9a* shows this applied to the fungal image. Such smoothing only rarely facilitates manual interpretation, although it can help automatic operations.

It is easy to define modifications to the above filters. We may wish to give more weight to the centre pixel (coefficient *e*) than to the others, or more weight to the horizontal and vertical neighbours than to the diagonal ones. Filters with coefficients

$$\begin{matrix} 1/10 & 1/10 & 1/10 \\ 1/10 & 2/10 & 1/10 \\ 1/10 & 1/10 & 1/10 \end{matrix} \quad \text{or} \quad \begin{matrix} 1/16 & 2/16 & 1/16 \\ 2/16 & 4/16 & 2/16 \\ 1/16 & 2/16 & 1/16 \end{matrix}$$

have thus been proposed. The second of these reduces the tendency of the simple moving average to smooth more along diagonals than horizontally or vertically. (In smoothing filters, the weights should be scaled to sum to one).

It can be useful to smooth in horizontal or vertical directions only, i.e. to use filters with coefficients proportional to

$$\begin{matrix} 0 & 0 & 0 \\ 1/3 & 1/3 & 1/3 \\ 0 & 0 & 0 \end{matrix} \quad \text{or} \quad \begin{matrix} 0 & 1/3 & 0 \\ 0 & 1/3 & 0 \\ 0 & 1/3 & 0 \end{matrix}$$

Repeating such a filter several times (say four or more) produces a filter with approximately Gaussian weights—a smooth bell-shaped pattern which tapers gradually to zero. It can be shown, given some assumptions about the nature of the noise and the aims of filtering, that it is optimal. If a Gaussian filter in the horizontal direction is followed by the same filter in the vertical direction, a circularly symmetric two-dimensional Gaussian filter results. (This follows from the fact that multiplying horizontal and vertical Gaussian functions, $\exp(-ax^2) \times \exp(-ay^2)$, produces a two-dimensional Gaussian function

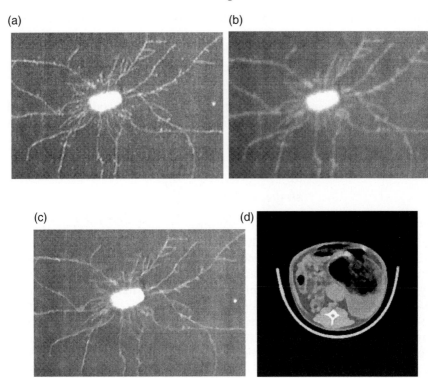

Figure 9. Smoothing filters. (a) Moving average in 3 × 3 window applied to fungal image. (b) Iteration of 3 × 3 moving average filter four times. (c) and (d) Median filter in 3 × 3 window iterated three times.

$\exp(-a(x^2 + y^2))$.) It provides an efficient way to calculate the two-dimensional filter. This can also be produced by repeated iteration of the simple filter with all weights equal. It is illustrated in *Figure 9b*. This is an example of the principle that smoothing filters, linear or otherwise, can be iterated to produce even smoother filters. Iteration effectively extends the filtering to a larger neighbourhood.

Sharpening filters

The extent to which a pixel differs from its background can be obtained from a filter which obtains the difference between a pixel value and the average of the eight neighbours:

$$\begin{array}{ccc} -1/9 & -1/9 & -1/9 \\ -1/9 & -8/9 & -1/9 \\ -1/9 & -1/9 & -1/9 \end{array}$$

If we then add this difference to the original image, we will have enhanced features such as edges. This corresponds to a filter proportional to

2: Biological image processing and enhancement

$$\begin{array}{ccc} -1/9 & -1/9 & -1/9 \\ -1/9 & 17/9 & -1/9 \\ -1/9 & -1/9 & -1/9 \end{array}$$

This is termed the unsharp masking filter. *Figure 10a* shows this filter applied to the CT image. If the image is noisy, this will also have been enhanced. This undesirable effect makes this filter unsuitable for such images.

Edge-detecting filters

Filters with weights proportional to

$$\begin{array}{ccc} -1 & 0 & 1 \\ -1 & 0 & 1 \\ -1 & 0 & 1 \end{array} \quad \text{or} \quad \begin{array}{ccc} -1 & -1 & -1 \\ 0 & 0 & 0 \\ 1 & 1 & 1 \end{array}$$

will give large responses only near vertical or horizontal edges, respectively. They are useful for transforming an image into one where the bright features correspond to edges in the original image. The vertical edge filter is illustrated in *Figure 10b*. These filters will respond to noise as well as edges, and so it may be useful to smooth the image before using them. Usually we are interested in edges in all directions, and so some combination of the above filters would be used. This is done in the Sobel filter, described below.

Fourier methods

Instead of representing an image as an array of pixel values, we can alternatively represent it as the sum of many sine waves of different frequencies, amplitudes, and directions. This is referred to as the frequency domain or Fourier representation. The reasons for taking this approach are because: extra insight can be gained into how linear filters work by studying them in the

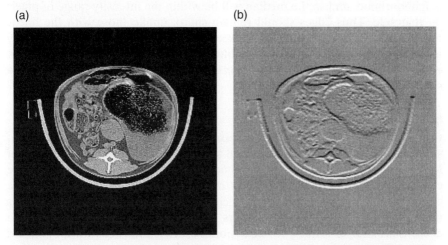

Figure 10. (a) Sharpening filter applied to fungal image (*Figure 3*). (b) Vertical edge filter.

frequency domain; some linear filters can be computed more efficiently in the frequency domain, by using the fast Fourier transform (FFT); and finally, new filters can be identified. (See for example ref. 4, Section 3.2.)

A linear filtering operation in the spatial domain is equivalent to a point-by-point multiplication of coefficients in the frequency domain. Therefore, an alternative way of applying a linear filter is to compute Fourier transforms of the image and the array of filter weights, multiply their Fourier coefficients and back-transform. Also, we can interpret filters by their effects on the Fourier coefficients of images. Smoothing filters, such as the moving average and Gaussian, are low-pass filters, because these filters only let through low-frequency terms, with coefficients at higher frequencies being set to zero. All high-frequency terms, corresponding to both noise and edges, are reduced to zero. In contrast, edge filters are bandpass filters because they remove both the lowest and highest frequencies from an image.

3.2 Non-linear filters

Linear filters are easy to use and well understood, but have the drawback of being rather limited. In particular, they are unable to reduce noise levels without simultaneously blurring edges. With non-linear filters, we can define any function of the pixels in a neighbourhood and noise reduction without blurring is possible. However, more care is needed in using them.

Median filters

The most common use of non-linear filters is to smooth without blurring edges. The median filter is the best known. It replaces the pixel with the median of the pixel intensities in a neighbourhood. For a pixel near an edge, pixels on the same side of the edge will be in a majority in a square or circular neighbourhood, and so the median will be within the intensity range of pixels on that side. Thus edges should remain sharp, unlike those with the linear moving average filters. This can be seen in *Figure*s *9c* and *9d*. As with any other smoothing filter, the median filter can be iterated. The result is not equivalent to a median filter in a larger neighbourhood, although it should be similar to it.

Spatially adaptive filters

Complex algorithms based on the pixels in a neighbourhood can be constructed to produce filters. One well-known class is based on the idea of looking at a number of subsets of the neighbourhood to determine which the centre pixel 'belongs to' or which is most uniform, and then replacing the centre pixel with some average of the chosen window. A review of such filters may be found in ref. 5. These filters can be quite powerful, although no consensus exists, or is likely to occur, about exactly when and how they should be used. The user should proceed with caution.

Edge filters

Perhaps no other topic in image analysis has attracted so much attention or led to such a huge literature and plethora of ideas as edge detection. Even though image analysis as a subject has been developing for about 30 years, new ideas for this fundamental task continue to be proposed. There is clearly no single best way to detect edges, and space here does not permit a review of even the most well-known approaches to edge detection. We mention just two. Both are non-linear functions of the output of a linear filter.

The *Sobel* filter is one of the most useful and widely available edge filters. It first estimates the intensity gradient in the horizontal and vertical directions from linear filters with coefficients

$$\begin{matrix} -1 & 0 & 1 \\ -2 & 0 & 2 \\ -1 & 0 & 1 \end{matrix} \quad \text{and} \quad \begin{matrix} -1 & -2 & -1 \\ 0 & 0 & 0 \\ -1 & -2 & -1 \end{matrix}$$

respectively. The maximum intensity gradient is then estimated as the square root of the sum of the squares of the horizontal and vertical gradients. The direction of the maximum gradient is also available from the arctangent of the ratio of the vertical and horizontal gradients. The Sobel filter is illustrated in *Figure 11a*. Stronger intensity gradients are shown as darker intensities.

The *Canny* filter (6) finds edge positions as maxima of the first derivative in the direction of steepest gradient, after a Gaussian filter. It is regarded as one of the best edge filters available, and has some mathematical properties of optimality. By incorporating a Gaussian filter, whose extent can be specified, we can smooth finer details and choose the scale at which we wish to find edges. The Canny filter is more complex than the Sobel filter, and produces edge decisions rather than edge evidence. It is illustrated in *Figure 11b*.

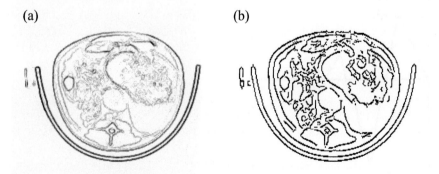

Figure 11. (a) Sobel filter applied to *Figure 3*. (b) Canny filter applied to *Figure 3* after four iterations of a 3 × 3 moving average filter.

Wavelets

Wavelets offer another approach to image smoothing. They are a recently developed method of signal and image representation with many possible uses, but here we focus on smoothing. The ideas and mathematics of wavelets are elegant, but subtle and complex. Introductions are given in refs 7–10.

The fundamental idea in wavelets is that we regard the image as being constructed as a sum of wavelet functions in different positions and at different scales—the multiresolution representation. The functions commonly used are surprisingly rough, but it can be shown that they can provide good approximations to smooth changes in the value of functions (in our case, image intensity). *Figure 12a* shows a plot of one such function, the Daubechies-6 wavelet. Two-dimensional versions are obtained by multiplying the function in the horizontal and vertical directions. We have shown no scale on the plot: a function is approximated by combinations of this wavelet function at different positions (translations) and scales (magnifications).

The coefficients of all wavelet functions at scalings differing by a factor of two and positions separated by integer numbers of pixels provide a complete representation of an image. However, we may believe that many of the finer scale wavelet coefficients reflect only noise in the image, and that if they were removed or reduced, the resulting image would be smoother. This idea has been developed into the theory of wavelet shrinkage (11). Wavelet coefficients are moved nearer to zero by an amount z, or set to zero if the absolute coefficient is less than z. Donoho *et al.* (11) consider the effects of this, and how to choose z.

Figure 12b shows the effect of this shrinkage on the CT image. We can see

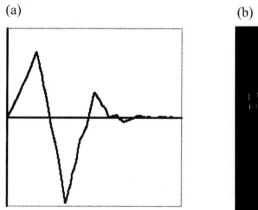
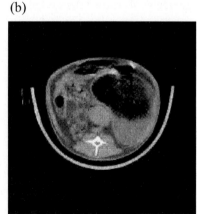

Figure 12. (a) The Daubechies-6 wavelet. (b) The CT image (*Figure 3*) smoothed by shrinking a Daubechies-6 representation at the lowest three resolution levels.

that some intensity variation within tissue types has been reduced, while real structures have mostly been unaffected.

Protocol 3. Smoothing a noisy image

We suggest the following approach to deciding how best to smooth a noisy image.

1. If the noise consists of fine speckle, try a median filter. It should be tried in a small neighbourhood first, and then for larger neighbourhoods. If it is not available in the software for larger neighbourhoods, try iterating the filter for smaller neighbourhoods. (Note that this not equivalent to using a larger neighbourhood, but should have a similar effect.) Always evaluate the effectiveness of the noise removal *and* the effect on features of interest.
2. If the speckle is mainly dark or light, try a morphological operation (see Section 4) such as a greylevel closing or opening with a cylinder or sphere respectively. These may also be termed maximum or minimum filters, respectively.
3. If you have a lot of noise, experiment with wavelet thresholding. The Daubechies-6 and Daubechies-8 wavelets are fairly effective and straightforward to use. Thresholds and resolution levels will need to be chosen.
4. If noise features are not all small-scale speckle, try alternating sequential morphological filters (Section 4). These may be encoded as a single step in software, or you may need to construct them yourself. Evaluate for a range of filter sizes.
5. If you intend to process a thresholded version of the image, try filtering before and after thresholding to determine which is most effective. Morphological filters for binary images are covered in the next section.
6. If none of the above are effective, it may be useful to try morphological operations which filter objects on the basis of size or shape criteria. These need to be used with care to avoid affecting features of interest.

---

4. Mathematical morphology

Mathematical morphology is an approach to image analysis based on set theory. This is unlike many other methods of image analysis, which derive from arithmetic. The seminal work on mathematical morphology is due to Serra and Matheron of the Ecole des Mines in Fontainebleau, France. The most comprehensive (and mathematically rigorous) reference is to be found in ref. 12. Other useful introductions may be found in refs 4 and 13–15.

4.1 Basic ideas

For binary images, sets are simply groups of pixels, and the terminology is just a convenient way of describing which pixels lie in particular groups. We present here a simplified view of the role of morphological operations in image analysis, looking only at their role in feature enhancement. Morphology involves probing the image with a structuring element. This is a set of pixels with a reference point. The simplest example, and the most commonly used, is a disc of radius r, with the reference point at the centre. A fundamental morphological operation is the erosion. The erosion of a set (group of pixels) by a structuring element is the set of pixel positions for which a structuring element placed with its reference point there will be contained completely within the set. In the case of the disc, a pixel is in the erosion of a set if a disc placed with its centre at that pixel is contained within the set, i.e. the pixel is further from the background than the radius of the disc. A dilation is an erosion of the background (the complement) of a set.

An opening is similar to an erosion, except that it consists of all points of the structuring element (not just the reference point) when the structuring element can be placed within a set. A closing is an opening of the background. These are illustrated in *Figure 13*. *Figure 13a* shows a subset of the fungal image, and *Figure 13b* the results of thresholding at a pixel intensity of 170. The effect of eroding and opening with a 3×3 square (an approximation to a small disc) is shown in *Figure*s *13c* and *13d* respectively. Erosions and openings remove pixels from a set, and the affected pixels are shown in grey. Dilation and closing add pixels to a set. These are illustrated in *Figure 13e* and *13f* respectively, with added pixels in grey. Here the closing is perhaps the most useful operation, in that it tends to close the gaps remaining after the thresholding. Opening would eliminate small or thin structures. *Figure 13g* shows the whole fungal image thresholded at 145 (i.e. *Figure 8e*), and closed with the 3×3 square. *Figure 13h* shows the result of closing with a disc of radius 4. Most of the gaps in the mycelia have been closed, although the effect near the centre of the image is less desirable.

Alternating sequential filters

If a lot of smoothing is needed, we may wish to use openings or closings with large structuring elements. However, these can be affected by fine detail in the noise. For example, a large disc may fail to fit within a larger object if the object contains even a single background pixel. We may be able to overcome this difficulty by alternating sequential filters. We start by opening and then closing with small structuring elements, and then continue to do this with structuring elements increasing in size, stopping when we feel that enough smoothing has been achieved. These filters are suitable when there are noise features at a range of scales. For images such as the fungal mycelia, where

2: Biological image processing and enhancement

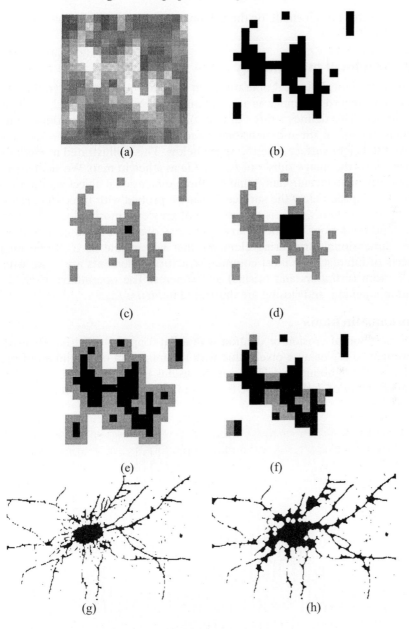

Figure 13. (a) A subset of the fungal image. (b) The result of thresholding at 170. (c) Erosion with a 3 × 3 square (an approximation to a small disc). The pixels removed are shown in grey. (d) Opening. (e) Dilation. Here the pixels added are shown in grey. (f) Closing. (g) The whole fungal image thresholded at 145 (i.e. *Figure 8e*), and closed with a 3 × 3 square. (h) The result of closing with a disc of radius 4.

noise features are all small, or the CT image, which has little noise, they are less useful.

4.2 Greylevel morphology

To perform morphological operations on an image of greylevels (rather than binary) we regard the image as a height map: the brighter the pixel, the higher the surface. This defines a three-dimensional object, and we can think of it as a binary image in three-dimensional space, which is divided into the space above the height surface, and the space below. This is illustrated in *Figure 14*, which shows the image subset of *Figure 13a* as a height map. We shall assume (arbitrarily but conventionally) that the three-dimensional object we study consists of the space below the surface. It can be probed with three-dimensional structuring elements. This is the essence of greylevel morphology. All of the ideas and definitions for the binary case carry over. We need only to devise three-dimensional structuring elements that will enable us to probe image aspects of interest. The most common structuring elements are those with a flat bottom (cylinders and prisms) and spheres. The operations of erosion, dilation, opening, and closing are illustrated in *Figure 15*.

Max and Min filters

The operation of erosion or dilation with a flat-topped structuring element is equivalent to replacing a pixel value with the minimum or maximum of pixel values in a neighbourhood. This can be useful for removing noise which is always brighter or darker than the features of interest. Using the corresponding

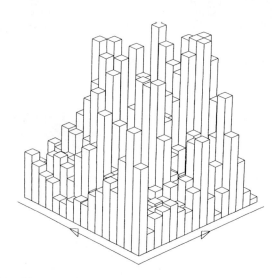

Figure 14. The fungal image subset shown in *Figure 13a* as a height map, viewed from the top right corner.

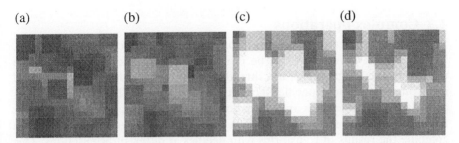

Figure 15. Greyscale morphological operations on *Figure 13a*, using a 3 × 3 cube: (a) erosion; (b) opening; (c) dilation; (d) closing.

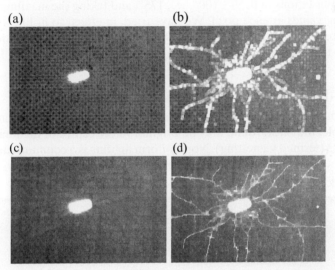

Figure 16. Minimum and maximum filters, in a disc of radius 3, applied to the fungal image. (a) Minimum (erosion); (b) maximum (dilation); (c) opening (max-after-min); (d) closing (min-after-max).

opening and closing (i.e. max-after-min or min-after-max filters) may achieve the same effect as erosion or dilation respectively, with less distortion of features of interest. This is illustrated in *Figure 16*. The minimum filter effectively removes everything except the central food source, while the max-after-min restores it to original size. The maximum filter expands the mycelia, while the min-after-max produces an image in which they appear more continuous.

The rolling ball transform

Morphological openings or closings with a sphere are termed rolling ball transforms. They produce results similar to the corresponding operations with a cylinder. They smooth a little less, and are better at preserving corners of objects.

4.3 Complex operations

We have only indicated some of the common morphological operations here. More complex procedures can be developed for specific applications. The mycelia in the fungal image are threadlike, and they could best be explored using one-dimensional 'needle-like' structuring elements. The difficulty is that a structuring element has an orientation: a needle oriented at $0°$ is different from one oriented at $90°$. The mycelia grow in many directions. We need to combine the results of transforms with needles at different orientations. This is not difficult to do, but substantial computation is required. *Figure 17* shows the result of performing openings with line segments (needles) of length 24 pixels at orientations of $0, 50°, 100°, \ldots, 175°$, and taking the maximum of the 36 resulting values at each pixel. We have tried, in effect, to fit line segments at 36 different orientations and chosen the best-fitting one. This transform has smoothed the mycelia and eliminated bright speckly features not lying on mycelia.

4.4 Top-hat transform and background subtraction

In many image formation systems there is variation in background intensity. For example, the background may be brighter in the centre of the image than at the edges (termed vignetting). Non-uniform lighting is a common cause, and there are others. This variation can interfere with image analysis procedures,

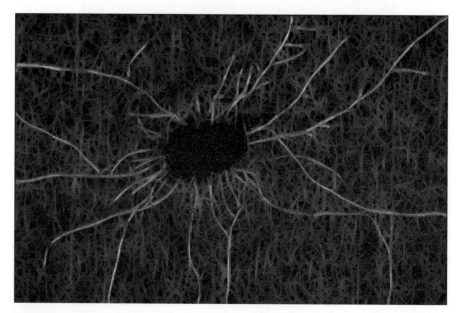

Figure 17. Maximum of openings with line segments of length 24 pixels at orientations of $0, 5°, 10°, \ldots, 175°$.

2: Biological image processing and enhancement

producing spurious artefacts or results which vary with position in the image. If so, it is desirable to reduce or eliminate it. Although this may not greatly affect manual interpretation, it can make automatic processing much easier.

Sometimes, if it is particularly severe, background intensity variation will be apparent when looking at the raw image. It may help to look at a blank image, i.e. one without a specimen. More often, some enhancement of the background intensity range is needed for it to become apparent. It may also be seen when thresholding the image with a threshold within the background intensity range. This can be seen in *Figure 8*. There is much more background noise on the right of the image than the left.

If it seems useful to reduce background variation, the first approach to this, remembering our first principle, is to try to improve the image formation. Establish the reason for the background intensity variation and correct or minimize it.

If significant background variation remains after the image formation has been made as good as possible, there are two approaches to an image-processing solution. Background subtraction is available in many image analysis packages. A blank image of the background only is recorded, and subtracted from subsequent images. This should leave only the specimen (see Chapter 1, Section 3.3).

Large-window morphological filters can eliminate the specimen from an image, leaving only the background, which may then be subtracted. This is known as the top-hat filter. We close with a cylinder (whence the term top-hat), then subtract the closing from the original image to find narrow dark features in the image. This is because the closing will have eliminated them, and they will be apparent when the closing is subtracted from the original image. For finding bright features, the subtraction of the opening is appropriate. This is illustrated for the fungal image in *Figure 18*.

Protocol 4. Background subtraction

1. Determine the extent to which variation in the background is a problem. This will depend on how much variation there is, and what image analysis is intended.
2. Try to reduce background variation in the image formation by ensuring uniform illumination. If it remains a problem, try either or both of the following steps.
3. Store an image of the background only, and subtract it from, or divide it into, all subsequent images. If the images are noisy, use the average of several background images.
4. Use a morphological filter such as the top-hat (Section 4.4) which filters out image objects to leave only the background, and then subtract this from the original.

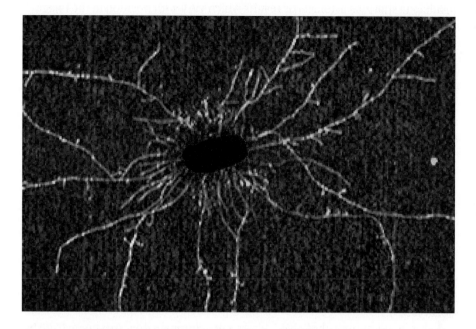

Figure 18. Top-hat filter applied for background removal in the fungal image.

5. Texture measures

Texture is an important characteristic in the analysis of many images. Despite being a property of almost all surfaces, no formal definition of texture exists. It is generally viewed as a measure of properties such as smoothness, coarseness, and regularity. Texture is a property of areas, the texture of a point being undefined. It is invariant to changes in orientation, brightness of the image, and size of the area. It is, however, dependent on resolution and magnification. There are essentially three main approaches used in analysing texture—the *statistical*, the *structural*, and the *modelling* approach.

In statistical texture analysis procedures a textured image is represented as a set of measurements called a feature vector. Many statistical methods are based on variations of co-occurrence (or greylevel spatial dependence) matrices. Other statistical approaches include the use of greylevel difference statistics and greylevel run-length matrices, methods based on texture edges and filter masks, and procedures using the power spectrum of the image.

Structural approaches try to characterize the pattern which is repeated in texture. They are therefore more appropriate for textures with fairly regular structures. They involve the definition of primitives, i.e. objects or patterns which are repeated in a given area. These primitives are extracted from an image and the texture is characterized by means of some rule that limits the number of possible arrangements of the primitive.

With the modelling approach texture is assumed to be a realization of a stochastic process which is governed by some parameters. Texture analysis is viewed as a parameter estimation problem: given a textured image the problem is to estimate the parameters of the assumed random process. The estimated parameters can then serve as features for texture classification and segmentation problems. Although several models have been used to generate and represent textures, the most common approaches are based on Markov random field models.

A review of recent developments in texture analysis is given in ref. 16.

5.1 Statistical approaches

Co-occurrence matrices

For an $n \times n$ image with pixel intensities ranging from 0 to 255, a 256×256 co-occurrence matrix is constructed by computing the number of times two pixels separated by distance d at a specified angle θ (displacement d, θ) occur in the image, one with intensity i and the other with intensity j. From this co-occurrence matrix a normalized co-occurrence matrix can be computed by dividing each element of the matrix by the number of possible pairs of neighbouring pixels. (Note that co-occurrence matrices are usually taken to be symmetric with pairs of greylevels at angle θ and $\theta + 180°$ being counted.) If a texture is coarse, and d is small compared to the size of the texture elements, the pairs of points at separation d should tend to have similar intensities. This means that the high values in the co-occurrence matrix will be concentrated on or near its main diagonal. Conversely, for a fine texture, if d is comparable to the texture element size, then the greylevels of points separated by d will be quite different, so that values in the co-occurrence matrix will be spread out relatively uniformly. Thus a good way to analyse texture coarseness is to compute some measure of the scatter of the co-occurrence matrix values around the main diagonal. Similarly, if the texture is directional then the degree of spread of the values about the main diagonal in the co-occurrence matrix should vary with the direction θ. Thus texture directionality can be analysed by comparing spread measures of the co-occurrence matrix for various values of θ. Haralick et al. (17) suggested a set of 14 textural features which can be extracted from co-occurrence matrices and which contain information about image textural characteristics such as homogeneity, linearity, and contrast. Since then, several other textural features have been proposed. Four of the more commonly used features are described here. In the following, $P(i,j)$ is the (i,j)th element of the given normalized co-occurrence matrix for a given displacement.

Angular second moment $\quad \Sigma\Sigma P(i,j)^2$
Contrast $\quad \Sigma(i-j)^2 \Sigma\Sigma P(i,j)$
Correlation $\quad \dfrac{\Sigma\Sigma[ijP(i,j)] - \mu_i\mu_j}{\{\sigma_i\sigma_j\}}$

Entropy $\quad -\Sigma\Sigma P(i,j)\ln P(i,j)$

where $\mu_i = \Sigma i P(i,j)$, $\sigma_i^2 = \Sigma i^2 P(i,j) - \mu_i^2$; μ_j and σ_j^2 are similarly defined. Angular second moment is a measure of homogeneity. A homogeneous image will result in a co-occurrence matrix with a combination of high and low $P(i,j)$s. In particular, where the range of greylevels is small, the $P(i,j)$s will tend to be clustered around the main diagonal resulting in a high value of angular second moment. A non-homogeneous image will result in an even spread of $P(i,j)$s and hence a low angular second moment. Contrast is a measure of the local variations present in an image. If there is a large amount of variation in an image the $P(i,j)$s will be concentrated away from the main diagonal and the contrast feature will be high. Correlation is a measure of greytone linear dependencies in the image and will be high if an image contains a considerable amount of linear structure. Entropy is another measure of the homogeneity of an image. It is large when all the $P(i,j)$s are of similar magnitude and small when the $P(i,j)$s are unequal.

Table 1 shows the value of the four features, for a (3,0) separation, for the subimages of *Figure 1d*. Good separation can be seen between A and B and between C and D on all features.

One problem with co-occurrence matrices is that they are dependent on the angle θ used to compute them. The most commonly used way of avoiding this problem is to compute co-occurrence matrices over four directions and calculate an average matrix. Another solution is to use neighbouring greylevel dependence matrices. For a given d and a, a neighbouring greylevel dependence matrix is computed by counting the number of times the difference in

Table 1. Texture summaries for bone subimages

	Subimage			
	A	B	C	D
Co-occurrence matrix summaries				
Angular second moment ($\times 10^3$)	4.63	3.60	2.59	2.52
Contrast	601	936	6428	6482
Correlation ($\times 10^5$)	−3.68	−2.60	−2.01	−2.17
Entropy	12.5	12.9	13.4	13.5
Difference statistics				
Angular second moment ($\times 10^2$)	301	468	3214	3241
Contrast	3.48	2.76	1.09	1.10
Entropy	3.54	3.77	4.67	4.67
Mean	13.1	16.7	44.8	45.7

pixel intensity between each element in the image is equal to or less than a at a certain distance d. As with co-occurrence matrices, several features can be computed from neighbouring greylevel dependence matrices to describe the texture.

Greylevel difference statistics

Greylevel difference statistics are based on absolute differences between pairs of greylevels. As with co-occurrence matrices, the starting point is an $n \times n$ image consisting of 256 greylevels. The greylevel difference statistics are contained in a 256-dimensional vector and are computed by taking the absolute differences of all possible pairs of greylevels distance d apart at angle θ and counting the number of times the difference is 0, 1, ..., 255. The difference statistics are then normalized by dividing each element of the vector by the number of possible pairs of pixels. Four measures can be computed from these vectors to describe the texture.

Angular second moment $\quad \Sigma P(i)^2$
Contrast $\quad \Sigma i^2 P(i)$
Entropy $\quad -\Sigma P(i) \ln P(i)$
Mean $\quad \dfrac{\Sigma i P(i)}{\text{LVL}}$

where the $P(i)$s are the ith elements of the vectors of normalized greylevel differences, and LVL is the number of greylevels.

Angular second moment, contrast, and entropy provide similar textural information to those with the same name based on co-occurrence matrices. Mean is essentially another measure of contrast and will be small when neighbouring pixels have similar greylevels and the $P(i)$ are concentrated near the top of the vector. *Table 1* shows the value of the above four features for the subimages of *Figure 1d*.

Run-length statistics

For an $n \times n$ image consisting of 256 pixel intensities and direction θ, run-length statistics are calculated by counting the number of runs of a given length (from 1 to n) for each greylevel. In a coarse texture it is expected that long runs will occur relatively often, whereas a fine texture will contain a higher proportion of short runs. Five measures are often calculated from run-length statistics. These include long run emphasis which will be large when there are lots of long runs of the same greylevel, and short run emphasis which will be large when there are lots of short runs. Grey level distribution will be large when runs are not evenly distributed over the different grey-levels.

Other statistical techniques, which are not described here, are those based on the Fourier power spectrum, relative extrema techniques, and methods using filter masks.

5.2 Structural approaches

Structural approaches assume that textures are made up of regions or primitives which appear in regular repetitive arrangements. They therefore tend to be more appropriate for synthetic textures. The first step in describing a texture is to identify the primitives. A primitive is a connected set of resolution cells characterized by a list of attributes. The simplest primitive is a single pixel with its intensity as its attribute. More commonly, primitives are sets of pixels all with the same or similar intensity or edge direction. Attributes may simply be the intensity of the region, although other possibilities are size, shape, and direction. Once the primitives and their attributes have been identified, the next step is to determine the placement rule, i.e. the spatial relationship between the primitives. Often, there is no well-defined spatial relationship. For classification purposes, the extracted primitives can be used as texture features and standard classification techniques can then be used. Such approaches are often termed structural-statistical approaches.

5.3 Modelling approaches

Texture modelling techniques involve constructing models to specify textures. The object is to capture the intrinsic character of the texture in a few parameters. The models can then be used to generate synthetic textures or to describe an observed texture. Texture models can be broadly classified into stochastic models such as those based on Markov and Gibbs random field models, and structural models, which specify the manner in which an image is generated but have no probabilistic description. Structural models include random walk models and random mosaic models. Early literature in texture modelling concentrated on methods for texture synthesis, but recently there has been more interest in using texture models to segment texture.

Markov random field and Gibbs models

Markov random field and Gibbs models are the most popular models used to describe texture. The origins of Gibbs distributions lie in physics and statistical mechanics. An image is assumed to consist of a square array of pixels or sites (i,j), $1 \leq i \leq n$, $1 \leq j \leq n$. These sites are numbered row by row from 1 to n^2 starting in the upper left. Each site in an image is assigned a 'colour', where colour may be pixel intensity in the texture-modelling context, or texture type in classification and segmentation contexts. The idea of a neighbourhood is central to Gibbs and Markov models and the probability of colour i at site t will depend on the neighbours of t. The probability is estimated from an energy function which has several parameters, the number depending on the number of neighbours a site is assumed to have. The main difficulty with Markov and Gibbs models is how to estimate these parameters. Several methods have been proposed but few can easily be used in images with more than two greylevels. Two such methods are Besag's coding scheme (18) and the maximum pseudolikelihood method.

5.4 Texture transforms

Texture is a property of a region, not of a single pixel. However, it can be useful to obtain a measure of texture in a moving neighbourhood around every pixel of an image. This could be a starting point for segmenting an image on the basis of texture differences. *Figures 19a* and *19b* show the angular second moment and contrast respectively, histogram-equalized, in 32 × 32 neighbourhoods around even-numbered pixels, odd-numbered pixels being omitted to reduce the heavy computational burden. Pixels beyond the boundaries were defined by reflection in the boundaries. The separation between the two bone types is clear. *Figure 19c* shows the results of thresholding *Figure 19a* with a manually chosen threshold.

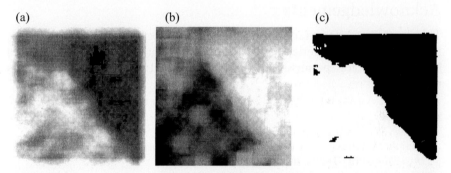

Figure 19. (a) Histogram-equalized angular second moment based on the co-occurrence matrix with displacement of (3,0), in a 32 × 32 neighbourhood around even-numbered pixels. (b) The co-occurrence matrix contrast, calculated as for (a). (c) Image (a) thresholded at 110.

Protocol 5. Measuring image texture

1. Does the image have a regular repeating pattern? If so, structural techniques may be appropriate. If not, statistical techniques should be used.

2. To classify or segment an image using texture, the size of the window in which to compute the texture features should be chosen so that windows not near texture boundaries should contain only one type. Large windows will take a long time to compute, so even if there are large homogeneous areas, it is probably best to use no larger than 64 × 64 pixel windows.

> **Protocol 5.** *Continued*
>
> 3. Co-occurrence and greylevel difference matrices are flexible and widely available tools for texture assessment.
>   - To determine the optimal displacement d, estimate the size of texture features in the images and choose d to be about the same size. Trial and error will be needed to determine the best size for classifying and segmenting images.
>   - Once the co-occurrence or difference matrices have been obtained, the calculation of feature summaries is quick and easy, and so several can be examined to see which best indicates texture differences of interest. Of the measures referred to in Section 5, the correlation best reflects greylevel linear dependencies, while the other measures summarize texture coarseness.

Acknowledgements

This work was supported by funds from the Scottish Office Agriculture, Environment and Fisheries Department. We are grateful to Lou Ralph and Mark Young for permission to use the fungal and X-ray CT images respectively.

References

1. Ralph, M.-L. (1997). *Factors affecting growth of the arbuscular mycorrhizal fungal mycelium*. PhD Thesis, Edinburgh University.
2. Gonzalez, R. C. and Wintz, P. (1987). *Digital image processing*, 2nd edition. Addison-Wesley, Reading, MA, USA.
3. Glasbey, C. A. (1993). *CVGIP: Graphical Models and Image Processing*, **55**, 532–537.
4. Glasbey, C. A. and Horgan, G. W. (1995). *Image analysis for the biological sciences*. Wiley, Chichester.
5. Chin, R. T. and Yeh, C. (1983). *Computer Vision, Graphics and Image Processing*, **23**, 67–91.
6. Canny, J. (1986). *IEEE Transactions on Pattern Analysis and Machine Intelligence*, **6**, 679–698.
7. Mallat, S. G. (1989). *IEEE Transactions on Pattern Analysis and Machine Intelligence*, **11**, 674–693.
8. Mallat, S. G. (1989). *Transactions of the American Mathematical Society*, **315**, 69–87.
9. Daubechies, I. (1992). *Ten lectures on wavelets*. Society for Industrial and Applied Mathematics, Philadelphia.
10. Kay, J. (1994). *1994: supplement to Vol. 21 of Applied Statistics*, **21**, 209–224.
11. Donoho, D. L., Johnstone, I. M., Kerkyacharian, G., and Picard, D. (1995). *Journal of the Royal Statistical Society*, B**57**, 301–337.

12. Serra, J. (1982). *Image analysis and mathematical morphology*. Academic Press, London.
13. Serra, J. (1986). *Computer Vision, Graphics and Image Processing*, **35**, 283–305.
14. Haralick, R. M., Sternberg, S. R., and Zhuang, X. (1987). *IEEE Transactions on Pattern Analysis and Machine Intelligence*, **9**, 532–550.
15. Haralick, R. M. and Shapiro, L. G. (1985). *Computer Vision, Graphics, and Image Processing*, **29**, 100–132.
16. Reed, T. R. and Dubuf, J. M. H. (1993). *CVGIP: Image Understanding*, **57**, 359–372.
17. Haralick, R. M., Shanmugam, K., and Dinstein, I. (1973). *IEEE Trans. Systems, Man and Cybernetics*, SMC-**3**, 6, 610–621.
18. Besag, J. E. (1974). *Journal of the Royal Statistical Society*, B**36**, 192–236.

3

Image analysis: quantitative interpretation of chromosome images

KENNETH R. CASTLEMAN

1. Introduction

In this chapter we illustrate how a set of image analysis techniques can be integrated to form a solution for a problem. One problem that illustrates this particularly well is chromosome analysis. Since the early 1960s, the field of cytogenetics has provided an important application area for digital image analysis techniques. Indeed, many of the methods now widely used in other fields were first applied to chromosomes. The object of automatic karyotyping is to recognize, in a microscope image, each of the 46 chromosomes from the nucleus of a human cell, and place them into a standardized arrangement called a *karyotype*. From this rearranged image the geneticist can make a diagnosis.

The development of image analysis techniques for automated chromosome karyotyping has a noble history that dates back to the 1960s (1). It was one of a handful of outstanding problems that stimulated early pattern recognition research. Numerous contributions have been made since then by researchers worldwide (2–34), including several pioneering developments in the field of pattern recognition research. Systems for automated karyotyping, commercially available since the mid-1980s, are now routinely used in cytogenetics laboratories. Over the last 15 years, commercial automated cytogenetics instruments have come into widespread usage for automatic karyotyping and anomaly detection of human and mammalian cells in clinical diagnosis and cancer research. The rapidly expanding workload in clinical and cancer cytogenetics laboratories precipitated this unprecedented rise in the use of cytogenetics automation equipment. Commercial instruments have reached the point where perhaps more than half the karyotypes produced in the world are done with computer assistance.

There are currently over 1200 clinical cytogenetics laboratories in the world using automated instrumentation to perform more than two million cytogenetics studies each year. In most of these, the primary factors of interest are cost per case and the number of cases that can be processed per day.

This chapter presents many of the quantitative image analysis and pattern recognition techniques that have proved useful in this application over the years.

1.1 The karyotyping problem

When a human cell goes into the metaphase stage of mitosis (cell division), the DNA in the nucleus condenses into 46 discrete packets called 'chromosomes'. Suitably prepared and stained, they take on a characteristic appearance, with four arms connected at the 'centromere' (*Figure 1a*). While the pattern is consistent in normal individuals, they exhibit specific morphological rearrangements in abnormals.

Prior to 1970 the chromosomes could be visually distinguished into seven groups, based on size and arm-length ratio. With the advent of staining techniques that cause them to exhibit a banding pattern, each of the 22 homologous chromosome pairs, and the two sex-determinant chromosomes, can be identified uniquely. Karyotyping is the process of placing the images of the chromosomes from a single cell into a stylized format that facilitates diagnosis (*Figure 1b*).

1.2 Historical perspective

Several groups have worked on the development of automatic chromosome classification techniques since James Butler's group applied it to marmoset chromosomes in the early 1960s (1). This development has led to today's common use of commercial automatic karyotyping systems in cytogenetics laboratories all over the world. Given the importance of this work, there remains continuing interest in improving the accuracy and throughput of these instruments.

1.3 Current practice

Today digital image analysis is used in commercially available systems, not only for automated karyotyping, but also for comparative genomic hybridization (CGH), fluorescent *in-situ* hybridization (FISH), and multicolour FISH (M-FISH). It has been employed for automatic metaphase finding and chromosome breakage detection as well.

1.4 Automatic karyotyping

The following sections describe automatic karyotyping as a four-stage process. The first stage is *image segmentation*, in which each chromosome is found and its image is isolated from the rest of the scene. The second stage is

Figure 1. Human chromosomes: (a) metaphase spread; (b) karyotype.

3: Image analysis: quantitative interpretation of chromosome images

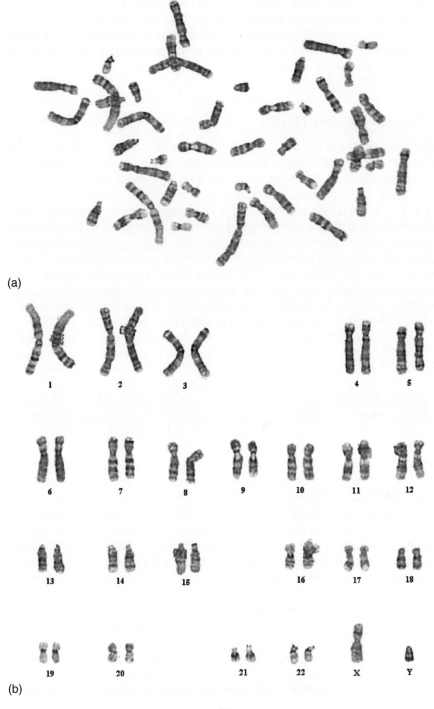

(a)

(b)

chromosome measurement, where quantitative values that characterize each chromosome are extracted from its image. The third stage is *chromosome classification*, the identification of each chromosome. The fourth stage is the actual generation and display of the final karyotype image, in which vertically oriented chromosomes are arrayed in a standard format.

2. Image segmentation

The *object locator* is the algorithm that isolates the images of the individual objects in the complex scene. The segmentation algorithm must laboriously isolate the objects in an image by breaking up that image into sets of pixels, each of which is the image of one object, a non-trivial task in digital image analysis. Since accurate segmentation of the image is vitally important, it may be done in as many as three steps. The first step establishes a preliminary boundary around each chromosome. This can be followed by a *boundary refinement* step which adjusts the initial boundaries to make them better fit the chromosomes. There may yet be segmentation errors—cases where single chromosomes have been broken apart or multiple chromosomes stuck together. This can be addressed in a third step that analyses each isolated object, seeking to detect and correct segmentation errors.

Connectivity. The image segmentation process partitions a digital image into disjoint (non-overlapping) regions. A *region* is a connected set of pixels. A connected set is one in which all the pixels are adjacent or touching. The formal definition of *connectedness* is as follows (35). Between any two pixels in a connected set, there exists a connected path wholly within the set. A *connected path* is one that always moves between neighbouring pixels. Thus, in a connected set, you can trace a connected path between any two pixels without ever leaving the set.

There are two rules of *connectivity*, and either one can be adopted (35). If only laterally adjacent pixels (up, down, right, left) are considered to be connected, this is '4-connectivity', and the objects are '4-connected'. Thus a pixel has only four neighbours to which it can be connected. If diagonally adjacent (45° neighbour) pixels are also considered to be connected, we have '8-connectivity', and the objects are '8-connected'. Each pixel has eight neighbours to which it can be connected. Either connectivity rule can be used as long as one is consistent. Often 8-connectivity yields results that align more closely with one's intuition.

Segmentation approaches. Image segmentation can be approached from three different philosophical perspectives. In what we call the *region approach*, one assigns each pixel to a particular object or region. In the *boundary approach*, one attempts to locate directly the boundaries that exist between the regions. In the *edge approach* one seeks first to identify edge pixels and then to link them together to form the required boundaries. All three

3: Image analysis: quantitative interpretation of chromosome images

approaches are useful for visualizing the problem and implementing a solution.

In this section, we examine several techniques for isolating the objects in an image. Once isolated, these objects can be measured and classified. Techniques for the latter two activities are addressed later in this chapter. Throughout this discussion we assume the chromosomes have high greylevel while the background has low greylevel.

2.1 Thresholding

Thresholding is a particularly useful technique for establishing boundaries in images which, like metaphases, contain solid objects resting on a contrasting background (36). It is computationally simple and never fails to define disjoint regions with closed, connected boundaries. Although those boundaries can be quite disappointing, they often can be improved by further processing.

When using a threshold rule for image segmentation, one first selects a threshold greylevel and then assigns all pixels below that greylevel to the background. All pixels with greylevel at or above the threshold are considered to be interior points. The different connected sets of interior points then constitute the separate objects in the image. The boundary is then the set of interior points each of which has at least one neighbour outside the object.

Thresholding works well if the objects of interest have reasonably uniform interior greylevel and rest on a background of unequal but relatively uniform greylevel. If the object differs from its background by some property other than greylevel (e.g. texture, etc.), one can first use an operation that converts that property to greylevel. Then greylevel thresholding can segment the processed image.

2.1.1 Fixed thresholding

In the simplest implementation of image segmentation by thresholding, the threshold greylevel value is held constant throughout the image. Properly selected, a fixed global threshold greylevel will usually work well when the background greylevel is reasonably constant throughout the image, and the objects all have approximately equal contrast with respect to the background.

2.1.2 Adaptive thresholding

In many cases, the background greylevel is not constant, and object contrast varies within the image. In such cases, a threshold that works well in one area might work poorly in other areas of the image. In these cases, it is convenient to use a threshold greylevel that is a slowly varying function of position in the image.

Figure 2a shows a microscope image of a fluorescently labelled metaphase. In this image, the background greylevel varies due to glare and non-uniform excitation, and brightness varies from one chromosome to the next. In *Figure 2b*, a constant threshold greylevel has been used throughout the image to

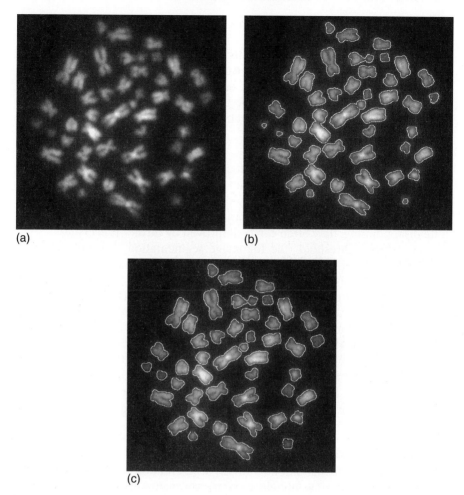

Figure 2. Adaptive thresholding: (a) fluorescently labelled metaphase spread; (b) segmentation with a fixed threshold; (c) segmentation with an adaptive threshold.

isolate the chromosomes. Note that the resulting boundaries are too small in the periphery of the spread, and too large towards the centre. There are four segmentation errors—cases where two chromosomes have been stuck together.

In *Figure 2c*, the threshold varied throughout the image, commensurate with local background and chromosome contrast. This adaptive segmentation technique was implemented as a five-step process. First the output of a Sobel edge detection operator (see Section 2.6.1, below) was thresholded to produce a mask showing where the edge pixels are located. Then 31-by-31-pixel neighbourhoods of the original image were averaged, using only pixels inside the mask, to determine the mean greylevel of edge points. This image was then smoothed by convolution with a lowpass filter. This produced a 'threshold map' that varied throughout the image, corresponding roughly to the local greylevel of edge pixels. This function was used as a greylevel threshold to segment the image. The example in *Figure 2c* shows only one segmentation error, and the remaining chromosomes, in general, are better represented by their boundaries.

2.1.3 Threshold selection

Unless the object in the image has extremely steep sides, the exact value of threshold greylevel can have considerable effect on the boundary position and thus the apparent size of the extracted object. This means that subsequent size measurements, particularly the area measurement, are rather sensitive to the threshold greylevel. For this reason, one needs an optimal, or at least consistent, method to establish the threshold.

An image containing an object on a contrasting background has a bimodal greylevel histogram. The two peaks correspond to the relatively large numbers of points inside and outside the object. The dip between the peaks corresponds to the relatively few points around the edge of the object. This dip is commonly used to establish the threshold greylevel (37–39).

The histogram is the derivative of the area function for an object whose boundary is defined by thresholding:

$$H(D) = - \frac{d}{dD} A(D) \qquad [1]$$

where D is the greylevel, $A(D)$ is the area of the object obtained by thresholding at greylevel D, and $H(D)$ is the histogram (36). Thus increasing the threshold from D to $D + \Delta D$ causes only a slight decrease in area if D is near the dip in the histogram. Therefore, placing the threshold at the dip in the histogram minimizes the sensitivity of the area measurement to small errors in threshold selection.

If the image containing the object is noisy and not large, the histogram itself will be noisy. Unless the dip is uncommonly sharp, the noise can make its location obscure, or at least unreliable, from one image to the next. This can be overcome to some extent by smoothing the histogram using either a convolution filter or a curve-fitting procedure. If the two peaks are of unequal size, severe smoothing will tend to shift the position of the minimum. The peaks, however, are rather easy to locate and relatively stable under reasonable amounts of smoothing. A more reliable method, then, is to place the threshold at some fixed position relative to the two peaks—perhaps midway (5). The two peaks represent the mode (most commonly occurring) greylevels of the interior and exterior points of the objects. It is often easier to estimate these parameters than the position of the dip in the histogram.

2.2 The watershed algorithm

A relative of adaptive thresholding is the watershed algorithm. Rather than simply thresholding the image at the optimum greylevel, the watershed approach begins with a threshold that is deliberately set too high, but that properly isolates the individual objects. This segments the image into the proper number of objects, but with boundaries that fit too tightly. Then the threshold is lowered gradually, one greylevel at a time, causing the object boundaries to expand. When they touch, however, the objects are not allowed to merge. These points of initial contact become the final boundaries separating adjacent objects. The process is terminated before the threshold reaches the greylevel of the background, that is, when the boundaries of all well-isolated objects are properly set. The final segmentation will be correct (i.e. one boundary per actual object in the image) if and only if correct segmentation is obtained at the starting threshold.

The watershed algorithm can be used with the distance transform (*Figure 3*). Even if the initial segmentation generates touching objects, the line of contact can still be identified. An auxiliary image is formed in which the value at each pixel is the distance to the boundary of the thresholded object (*Figure 3b*). The distance-transformed image can be thresholded at successively lower levels until the detected regions touch, as in the watershed algorithm (*Figure 3c*). This can be useful if the touching objects are roughly the same size, and their greylevels are not very uniform, leading to errors in the basic watershed algorithm.

2.3 Gradient image thresholding

One can compute the gradient magnitude of an image, $f(x,y)$, as

$$|\nabla(f(x,y))|^2 = \left(\frac{\partial f}{\partial x}\right)^2 + \left(\frac{\partial f}{\partial y}\right)^2 \qquad [2]$$

In a gradient image, both object and background pixels have low gradient magnitude, while edge points take on larger values. Kirsch's segmentation method (40) uses the watershed algorithm to exploit this phenomenon. One first thresholds the gradient image at a moderately low level, where the objects and the background are separated by bands of edge points that exceed the threshold. As the threshold is gradually increased, both the objects and the background grow in size. When they touch, they are not allowed to merge, but rather their contact points define the boundary.

While this method is more computationally expensive than thresholding, it produces maximum gradient boundaries, and these tend to fall where the human eye places boundaries. For multiple object images, the segmentation will be correct if and only if it is done correctly by the initial thresholding step. Smoothing the gradient image before thresholding produces smoother boundaries.

3: Image analysis: quantitative interpretation of chromosome images

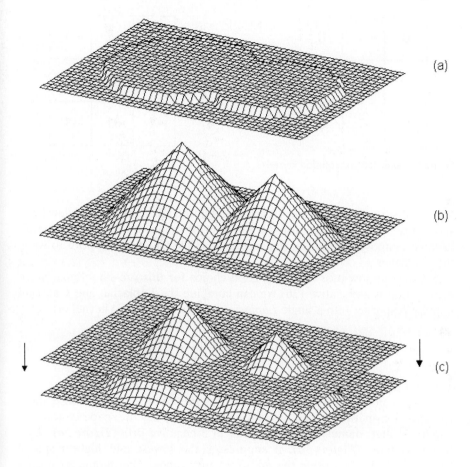

Figure 3. The distance transform—watershed algorithm: (a) binary image; (b) distance transform; (c) thresholding the distance transform.

2.5 Laplacian edge detection

The Laplacian is a scalar second derivative operator for functions of two dimensions, given by

$$\nabla^2 f(x,y) = \frac{\partial^2}{\partial x^2} f(x,y) + \frac{\partial^2}{\partial y^2} f(x,y) \qquad [3]$$

It can be implemented digitally by either of the convolution kernels in *Figure 4*.

As a second derivative, the Laplacian produces an abrupt zero crossing at an edge. Since it is a differentiation operator, a Laplacian filtered image will have zero mean greylevel.

The Laplacian can find the edges in a noise-free image. The binary image

0	−1	0
−1	4	−1
0	−1	0

−1	−1	−1
−1	8	−1
−1	−1	−1

Figure 4. Laplacian convolution kernels.

that results from thresholding a Laplacian filtered image at zero greylevel will show closed connected contours when interior points are eliminated. Noise, however, imposes a practical requirement for low-pass filtering prior to using the Laplacian.

A Gaussian low-pass filter is a good choice for this pre-smoothing. Since convolution is associative (36) we can combine the Laplacian and Gaussian impulse responses into a single *Laplacian of Gaussian* (LoG) kernel (41, 42):

$$-\nabla^2 \frac{1}{2\pi\sigma^2} e^{\frac{x^2+y^2}{2\sigma^2}} = \frac{1}{\pi\sigma^4}\left[1 - \frac{x^2+y^2}{2\sigma^2}\right]e^{\frac{x^2+y^2}{2\sigma^2}} \qquad [4]$$

This impulse response is separable in x and y, and thus the convolution operation can be implemented efficiently as the cascade of a row-wise and a column-wise operation. The LoG impulse response has the shape typical of a bandpass filter, namely a positive peak in a negative dish (*Figure 5a*). As a bandpass filter (*Figure 5b*), it suppresses the lowest and highest spatial frequencies while boosting the mid-frequency range. The parameter σ controls the width of the central peak and, thus, the amount of smoothing. The LoG filter is well approximated by a filter kernel that is the difference of two Gaussians of standard deviations σ_1 and σ_2 (36) when their ratio is $\sigma_2 = 1.6\sigma_1$ (41).

2.6 Edge detection and linking

Another approach to establishing the boundaries of the objects in an image is edge detection. Here one first examines each pixel, and its immediate neighbourhood, to determine if it is on the boundary of an object. Pixels that exhibit the required characteristics are labelled as '*edge points*'. An image in which edge points have been labelled normally shows each object outlined in edge points, but generally not with closed connected boundaries. Thus, a second step, '*edge point linking*', is required. This is the process of tying together nearby edge points to create closed connected boundaries.

3: Image analysis: quantitative interpretation of chromosome images

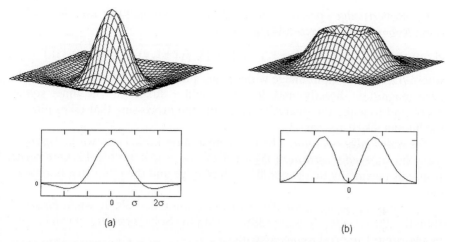

Figure 5. The Laplacian of a Gaussian filter: (a) impulse response; (b) transfer function.

2.6.1 Edge detection

If a pixel falls on the boundary of an object, its neighbourhood is a zone of greylevel transition. The two characteristics of principal interest are the slope and direction of that transition, that is the magnitude and direction of the gradient vector. Edge detection operators examine each pixel neighbourhood and quantify the slope, and often the direction as well, of the local greylevel transition. There operators are often convolution kernels that implement directional derivatives.

Protocol 1. Algorithm for edge detection using edge operators (convolution kernels)

A. *Convolve the image with a set of edge detection kernels*
1. Select the set of edge operators to be used (see choices below).
2. Convolve the image with each of the operators.

B. *Determine the edge magnitude and direction*
1. At each pixel, compare the magnitude of the output of each operator.
2. Take the largest output (or the square root of the sum of squares (43)) as the edge magnitude.
3. Take the direction of the largest output as the direction of the edge.
4. Repeat steps B1–B3 at every pixel in the image.
5. The result is an edge magnitude image and an edge direction image.

The Roberts edge operator. One local differential operator for edge finding is the *Roberts edge detector* (44), given by

$$g(x,y) = \{[\sqrt{f(x,y)} - \sqrt{f(x+1,y+1)}]^2 + [\sqrt{f(x+1,y)} - \sqrt{f(x,y+1)}]^2\}^{1/2} \quad [5]$$

where $f(x,y)$ is the input image with integer pixel coordinates x,y. It yields the edge magnitude directly, and does not yield a direction. The inner square roots tend to make the operation resemble the processing that takes place in the human visual system.

The Sobel edge operator. The two convolution kernels shown in *Figure 6a* form the *Sobel edge operator* (45) which can be used in protocol 1. One kernel responds maximally to a generally vertical edge and the other to a horizontal edge.

The Prewitt edge operator. The two convolution kernels shown in *Figure 6b* form the *Prewitt edge operator* (38). As with the Sobel operator, it can be used in algorithm 1 to produce an edge image.

The Kirsch edge operator. The eight convolution kernels shown in *Figure 6c*

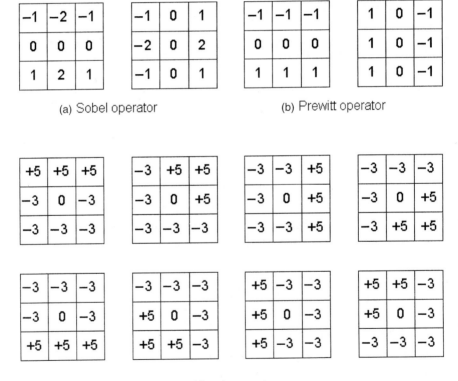

Figure 6. Edge detection operators: (a) Sobel; (b) Prewitt; (c) Kirsch.

3: Image analysis: quantitative interpretation of chromosome images

comprise the *Kirsch edge operator* (40). They respond maximally to edges oriented in directions spaced 45° apart. When used in algorithm 1 this operator produces finer direction information.

Edge detector performance. Visually, the edge images produced by these four edge operators usually appear rather similar, resembling a line drawing made from the picture. The Roberts operator, being 2 × 2, responds best to sharp transitions in low-noise images. The other three, being 3 × 3 operators, better handle more gradual transitions and noisier images. Experimentation is often needed to guide a selection for a particular type of image.

Multiresolution edge detection. Images often contain small, sharp edges as well as larger, more gradual changes of brightness. Larger edge detection operators will respond to larger edges, and vise versa. While it is possible to implement edge detectors larger than 3 × 3, the Canny edge detector applies a small kernel to a series of reduced-size images to obtain the same effect (46). This way both large and small edges are detected.

2.6.2 Edge linking

If the edges are reliably strong, and the noise level is low, one can threshold an edge magnitude image and thin the resulting binary image down to single-pixel-wide closed, connected boundaries. Under less than ideal conditions, however, such an edge image will have gaps that must be filled. Small gaps can be filled simply by searching a 5 × 5 or larger neighbourhood, centred on an endpoint, for other endpoints, and then filling in boundary pixels as required to connect them. In images with many edge points, however, this can over-segment the image. The following technique can overcome this problem.

Heuristic search. Suppose we have what appears to be a gap in a boundary in an edge image, but it is too long to fill accurately with a straight line, it may not really be a gap in the same boundary, or perhaps both. We can establish, as a quality measure, a function that can be computed for every connected path between the two endpoints, which we call A and B. This edge quality function could include the average of the edge strengths of the points, perhaps less some measure of their average disagreement in orientation angles (44, 47, 48).

Protocol 2. Heuristic search algorithm

Suppose we have what appears to be a gap in a boundary in an edge image between two particular edge points A and B.

A. *Evaluate neighbours of A as candidates for the first step towards B*
1. Identify the three neighbours of A that lie in the general direction of B.
2. Compute the edge quality function from A to each point from step 1.
3. Select the one that maximizes the edge quality function from A to that point.

Protocol 2. *Continued*

4. Use the point from step 2 as the starting point for the next iteration.
5. Repeat this process until point B is reached.

B. *Qualify the path*
1. Compare the edge quality function over the newly created path to a threshold.
2. If the newly created edge is not sufficiently strong, discard it.
3. If the newly created edge is sufficiently strong, take the resulting graph as the boundary.

Heuristic search techniques become computationally expensive if the edge quality function is complex and the gaps to be evaluated are many and long. It performs well in relatively simple images, but it does not necessarily converge upon the globally optimal path between endpoints.

Curve fitting. If the edge points are generally sparse, it might be desirable to fit a piecewise linear or higher order spline curve through them to establish a suitable object boundary. A simple example is the piecewise linear method called 'iterative endpoint fitting' (49).

Protocol 3. Iterative endpoint fitting algorithm

Suppose there is a group of edge points lying scattered between two particular edge points A and B, and we wish to select a subset of these to form the nodes of a piecewise linear path from A to B.

A. *Linking edge points*
1. Begin by establishing a straight line from A to B.
2. Compute the perpendicular distance from that line to each of the remaining edge points.
3. Identify the furthermost one, which becomes the next node on the path. The path now has two branches.
4. Repeat this process on each new branch of the path until no remaining edge point lies more that some fixed distance away from the nearest branch.

B. *Finding the boundary*
1. Repeat A for pairs of points A,B all around the object.
2. Take the resulting graph as a polygonal approximation to the boundary.

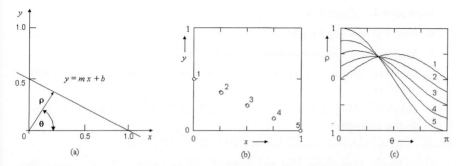

Figure 7. The Hough transform: (a) polar representation of a line; (b) the x,y plane; (c) the ρ,θ plane.

The Hough transform. The straight line $y = mx + b$ can be expressed in polar coordinates as

$$\rho = x\cos(\theta) + y\sin(\theta) \qquad [6]$$

(49) where ρ,θ defines a vector from the origin to the nearest point on that line (*Figure 7a*). This vector will be perpendicular to the line. Any line in the x,y plane corresponds to a point in the two-dimensional space defined by the parameters ρ and θ. Thus the Hough transform of a straight line in x,y space is a single point in ρ,θ space. Every straight line that passes through a particular point, x_1,y_1, in the x,y plane plots to a point in ρ,θ space, and these points must all satisfy *Equation 6* with x_1 and y_1 as constants. Thus the locus of all such lines in x,y space is a sinusoid in parameter space (*Figure 7c*), and any point in the x,y plane corresponds to a particular sinusoidal curve in ρ,θ space.

Suppose we have a set of edge points x_i,y_i that lie along a straight line having parameters ρ_o,θ_o (*Figure 7b*). Each edge point plots to a sinusoidal curve in ρ,θ space, but these curves must intersect at the point ρ_o,θ_o since this is a line they all have in common (*Figure 7c*). Algorithm 4 uses this to find the lines in an image.

Protocol 4. Piecewise linear boundary finding with the Hough transform

A. *Accumulate a two-dimensional histogram in ρ,θ space*

1. For each edge point, x_i,y_i, increment all the histogram bins in ρ,θ space that correspond to the Hough transform (sinusoidal curve) for a point at that location.
2. Repeat step 1 for all the edge points,

Protocol 4. *Continued*

B. *Find the line segments*
1. Search the ρ versus θ histogram for local maxima.
2. Generate lines corresponding to the ρ,θ coordinates of the local maxima.
3. Take the polygon formed by the resulting linear boundary segments as the boundary.

The Hough transform can be defined for geometric shapes other than straight lines. The circular Hough transform, for example, parameterizes a circle by its radius and the *x,y* coordinates of its centre point. Its output is thus a three-dimensional histogram. The preceding approach can be used to group together edge points that fall on the same circle. If the radius of the objects is known, it can be fixed beforehand, and only a two-dimensional (*x* versus *y*) histogram space must be searched, as with the linear Hough transform.

2.7 Region growing

The *region growing* approach (50–53) to image segmentation has proved useful in the field of computer vision.

Protocol 5. Region growing

A. *Prepare the image*
1. Determine a set of parameters whose values are different for different types of objects (including background as an object type).
2. Properties that distinguish interior and exterior pixels include average greylevel, texture, and colour information.
3. Divide the image into a large number of regions. These may be either small neighbourhoods or single pixels.

B. *Grow the regions*
1. In each region, compute the properties (from step A1) whose magnitudes reflect which type of object that region is in.
2. For each boundary between adjacent regions, compute a measure of boundary strength based on the differences in the properties of the two regions.
3. Compare the boundary strength to a predetermined threshold value.
4. If the boundary is strong, allow it to stand.
5. If the boundary is weak, dissolve it and merge the two regions into one.

> 6. Iterate the process by alternately recomputing region properties for the enlarged regions, computing strengths of all boundaries, and dissolving weak boundaries.
> 7. When no further boundaries are weak enough to be dissolved, the segmentation is complete.

Following the progress of this algorithm visually gives the impression of regions inside the objects growing until their boundaries coincide with the object edges. Region growing can utilize several image properties directly and simultaneously in determining the final boundary location, but it is computationally expensive. It has found less usage in chromosome images than in the segmentation of natural scenes, where strong a priori knowledge is unavailable.

3. Boundary refinement

An image segmentation operation produces a binary image (i.e. having only two greylevels). If the initial segmentation is less than satisfactory, then some form of subsequent processing can often improve the situation. We now examine methods for refining the boundary initially assigned to a chromosome.

3.1 Active contours

A promising image analysis tool for refinement of an object boundary is the *active contour* or 'snake' (54, 55). The curve formed by the connected edge points delineates the active contour. It is a set of connected points which iteratively move so as to minimize a specified energy function. Local properties of the image (e.g. greylevel, gradient, etc.) contribute to the energy of the snake, as do constraints on the continuity and curvature of the contour itself. In this way, the snake reacts to the image, but moves in a smooth, continuous manner towards the desired object boundary.

The so-called 'greedy' algorithm (56) for iterative boundary refinement is a simple and effective one.

> **Protocol 6.** The greedy snake algorithm
>
> A. *Prepare the image*
> 1. Define an energy function which would naturally take on low values for points located on the actual boundary.
> 2. Generate an initial boundary by a segmentation technique such as thresholding.

> **Protocol 6.** *Continued*
>
> **B.** *Wiggle the snake*
>
> 1. Compare the energy of each point on the boundary with the energies calculated for points in its neighbourhood.
> 2. For each current boundary point, move the boundary to the neighbouring point that has the lowest energy.
> 3. Perform this operation once on all points on the boundary to complete one iteration of the snake algorithm.
> 4. Repeat this iteration until it causes no further movement of the boundary points.
> 5. When this happens, the snake is said to have 'converged', and the segmentation is complete.

The choice of the energy function to be minimized determines the behaviour of the algorithm. Given a parametric representation of the snake, $v(s) = (x(s), y(s))$, where $s \in [0,1]$, the energy function

$$E^*_{snake} = \int_0^1 [E_{snake}(v(s))]ds = \int_0^1 [E_{int}(v(s)) + E_{image}(v(s)) + E_{ext}(v(s))]ds \quad [7]$$

is simply an integration of energy along the length of the contour.

The three energy terms in *Equation 7* correspond to the three 'forces' that act on the snake as it moves: (1) *internal forces* between the points of the contour (analogous to tension in and rigidity of the snake), (2) *image forces* (such as greylevel and gradient magnitude) that pull the snake towards the optimal boundary position, and (3) *external constraints* such as user forces applied to the contour (e.g. pulling on the contour with a virtual string).

The internal energy of the snake can be modelled using two terms (56): a continuity term (tension) and a curvature term (rigidity):

$$E_{int} = E_{cont} + E_{curve} \quad [8]$$

The continuity term can be made proportional to the difference between the current interpoint distance and the average interpoint distance for the entire snake. This formulation encourages equal spacing of contour points around the boundary. The energy associated with the curvature at a point can be approximated by taking the square of the magnitude of the difference between two adjacent unit tangent vectors. This gives a quick but reasonable estimate of local curvature. It tends to straighten the contour, thus favouring smoother outlines. The contributing measurement for the image energy term can be simply the result of an edge operator such as the Sobel gradient operator. The main advantages of the greedy algorithm are computational efficiency and its relative simplicity. Its main disadvantage is the extremely local nature of its decision criteria.

Figure 8 illustrates the progression of three active contours as they conform

3: Image analysis: quantitative interpretation of chromosome images

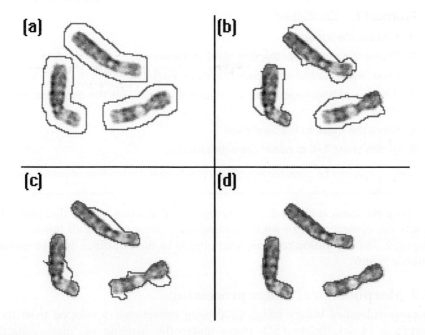

Figure 8. Active contour boundary refinement: (a) starting boundaries; (b) and (c) intermediate boundaries; (d) final boundaries.

to chromosome boundaries, starting from an initial boundary that is too large. Clearly, reasonable boundary initialization is necessary in the use of active contours, but the boundaries that result from well-configured snakes tend to agree with those assigned by the human eye, and to produce consistent measurement data.

3.2 Binary image processing

Locally implemented logical operations can implement boundary refinement. By definition, a boundary point is a pixel located inside an object, but which has at least one neighbour outside the object. One valuable tool is a 3 × 3 logical operation called a 'hit-or-miss transform'.

Protocol 7. The hit or miss transform

A. *Specify the mask*

1. Define a 3 × 3 (or larger) pattern of bits (mask) to be searched for in the image (e.g. all nine pixels black).
2. Define the output value for a 'hit' (0, 1, or unchanged).
3. Define the output value for a 'miss' (0, 1, or unchanged).

Protocol 7. *Continued*

B. *Process the image*
1. Centre the mask over the first pixel in the image.
2. If the neighbourhood matches the mask, set the output to the 'hit' value.
3. If the neighbourhood does not match the mask, set the output to the 'miss' value.
4. Move the mask to the next pixel.
5. Iterate steps 1–4 to cover the entire image.

Note: it may not be possible to process pixels near the borders of the image.

Suppose the mask is all black, for example, so that when it 'hits' that pattern, it sets the central pixel to white, leaving the central pixel of all 'misses' unchanged. This operation reduces solid objects to their outlines by eliminating interior points.

3.3 Morphological image processing

A powerful set of binary image processing operations, developed from the approach of set theory (57), come under the heading of 'mathematical morphology' (58–63). These simple basic operations can be concatenated to produce complex effects (64, 65).

Protocol 8. A morphological operation

A. *Specify the operation*
1. Define a 3 × 3 (or larger) pattern of bits (the 'structuring element') to be used in the operation (e.g. all nine pixels black).
2. Specify a logical operation to be performed between the structuring element and the underlying binary image.

B. *Process the image*
1. Centre the structuring element over the first pixel in the image.
2. Perform the specified logical operation between the structuring element and the underlying binary image.
3. Store the binary result of the logical operation at the corresponding location in the output image.
4. Move the structuring element to the next pixel.
5. Iterate steps 1–4 to cover the entire image.

Note: It may not be possible to process pixels near the borders of the image.

Similar to convolution, morphological image processing passes a fixed pattern over the image. Like the convolution kernel, the structuring element can be of any size, containing any complement of ones and zeros. The effect created depends upon the size and content of the structuring element and upon the nature of the logical operation. In the simple but important case of the 'basic' structuring element (which is 3 × 3, containing all ones) the choice of logical operation determines the outcome.

3.3.1 Erosion and dilation

Simple erosion is the process of eliminating all the boundary points from an object, leaving it smaller in area by one pixel all around the perimeter. The diameter of a circular object decreases by two pixels with each erosion. If it is less than three pixels thick at any point, it will be disconnected (into two objects) at that point. Objects no more than two pixels thick in any direction are eliminated. Erosion is useful for removing from a segmented image, objects that are too small to be of interest. The erosion operation, repeatedly applied, will shrink an object out of existence. Erosion results if the logical 'AND' operation is used with the basic 3 × 3 structuring element.

Dilation is simply the process of incorporating into the object all the background points that touch the object, leaving it larger in area by that amount. If the object is circular, its diameter increases by two pixels with each dilation. If two objects are separated by less than three pixels, they will become connected (merged into one object). Dilation is useful for filling holes in segmented objects. Dilation, repeatedly applied, will merge all the objects into one. Dilation results if the logical 'OR' operation is used with the basic 3 × 3 structuring element.

3.3.2 Opening and closing

The process of erosion followed by dilation is called 'opening'. It has the effect of eliminating small and thin objects, breaking objects at thin points, and generally smoothing the boundaries of larger objects, without significantly changing their area.

The process of dilation followed by erosion is called 'closing'. It has the effect of filling small and thin holes in objects, connecting nearby objects, and generally smoothing the boundaries of objects, without significantly changing their area.

Often when noisy images are segmented by thresholding, the resulting boundaries are quite ragged, the objects have false holes, and the background is peppered with small noise objects. Successive application of opening or closing operations can improve the situation markedly. Sometimes several iterations of erosion, followed by the same number of dilations, will produce the desired effect.

3.3.3 Thinning and skeletonization

Erosion can be implemented in such a way that single-pixel objects are left intact. This process, called 'shrinking', is useful when the total object count must be preserved. The first step of this two-step process is a conditional erosion in which pixels are marked as candidates for removal rather than actually being eliminated. In the second pass, those candidates that can be removed without destroying connectivity are eliminated while those that cannot are retained. 'Thinning' is the process of repeatedly shrinking an object till no further points can be removed. This reduces a curvilinear object to a single-pixel-wide line, showing graphically its topology.

In *Figure 9*, thinning a group of touching chromosomes produces a graph with typically one segment for each chromosome. This can form the basis for an algorithm that separates touching objects. Thinning can also be used iteratively to count the number of objects in a binary image. It is run repeatedly until all objects have shrunk to single points. Then the number of 'ones' remaining in the image is the object count.

A related operation is 'skeletonization', also known as the 'medial axis transform' or the 'grassfire technique' (66–70). The medial axis is the locus of the centres of all circles that are tangent to the boundary of the object at two or more disjoint points, but it is seldom implemented by actually fitting circles inside the object. To conceptualize the medial axis transform, imagine that a patch of grass, in the shape of the object, is set afire all around the periphery at once. As the fire progresses inwards, the locus of points where advancing fire lines meet is the medial axis. Skeletonization can be implemented with a two-pass conditional erosion, but the rule for pixel deletion is slightly different from that of thinning. The primary difference is that the medial axis extends to the boundary at corners, while the skeleton obtained by thinning does not.

Pruning. The thinning or skeletonizing process can leave short 'spurs' on the resulting figure. Since these are short branches, they are characterized by an endpoint that is located near an intersection. Spurs can be removed by first using a series of operations that remove only endpoints, thereby shortening all the branches, until the spurs disappear. This is followed by use of the original image to guide reconstruction of the branches that still exist.

Thickening. Some segmentation techniques tend to establish rather tight-fitting boundaries around objects in order to avoid merging them in error. Often the best boundary for object isolation is too tight for subsequent measurement. Thickening can correct this by enlarging the boundaries without merging nearby objects. Thickening can be done by dilation implemented in two passes, as with thinning. An alternative is to complement the image and use the thinning operation on the background. In fact, each type of erosion has a companion dilation-type operation when run on the background of a complemented image.

3: Image analysis: quantitative interpretation of chromosome images

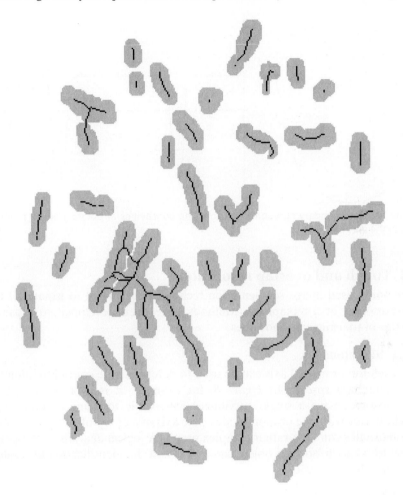

Figure 9. Chromosome thinning: chromosome outlines and medial axes.

3.4 Boundary curvature analysis

The curvature at a point on a curve is defined as the rate of change of the tangent angle at that point, as one traverses the curve. The curvature of an object's boundary is positive in regions where the object is convex and negative where it is concave. In *Figure 10*, for example, a plot of boundary curvature shows two sharp negative peaks corresponding to the two concavities. If the objects are expected to be convex, this signals a segmentation error. A cutting line, drawn between the two points, a and b, separates the two objects. Thus the boundary curvature function can assist the automatic detection and correction of segmentation errors.

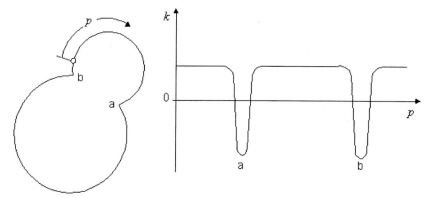

Figure 10. The boundary curvature function. The concavities at points a and b produce dips in the curve.

3.5 Touch and overlap resolution

Commonly used image segmentation techniques often fail to isolate all the chromosomes in a metaphase. Methods have been developed to separate clusters of touching chromosomes.

3.5.1. Skeletons

The skeleton or medial axis can be used to determine the overall topology of the metaphase spread. In *Figure 8*, for example, it is clear that isolated chromosomes have a single curvilinear axis, while touching chromosomes produce axes with one or more nodes. The axis of a group of several touching chromosomes can be a rather complex graph. A logical analysis of the graph structure of each isolated object can assist both the identification of clusters and their disaggregation.

3.5.2. Convex hulls

A convex object is one in which all lines between its boundary points fall inside the object. The convex hull of an object in a binary image is the convex shape one would obtain by stretching a rubber band around the object. It is the smallest convex polygon that contains the original object, and each vertex is a boundary point. It can be no smaller in area than the original object. Comparing the shape of the convex hull to that of the object itself can help determine the topology of the cluster. For example, for severely bent or touching chromosomes, the area of the hull is often significantly larger than that of the object itself.

The two classic methods of finding the convex hull of a set of points are Graham's Scan and Jarvis March. The latter, illustrated below, is sometimes called the 'package wrapping algorithm'. It walks around the object in a counterclockwise direction.

3: Image analysis: quantitative interpretation of chromosome images

---

Protocol 9. Convex hull generation by the Jarvis March algorithm

A. *Select the starting point*
1. Discard all interior points from the search.
2. Label the boundary point having the minimum *y*-coordinate as the starting point, 'O'.
3. If two or more fall on the same *y*-coordinate, take the one with the largest *x*-coordinate.
4. Label the starting point as 'A'.

B. *Compute edge angles*
1. Construct a line between A and another boundary point, B.
2. Compute the angle the line from A to B makes with the *x*-axis.
3. Repeat steps 1 and 2 for all remaining boundary points.
4. Label the boundary point that minimizes the angle as the next vertex on the convex hull.

C. *Iterate on the vertices*
1. Label the new vertex as point A.
2. Repeat steps B1–B4.
3. Continue steps C1 and C2 until point B has encircled the object and the next vertex is the starting point, O.
4. Take the resulting polygon as the convex hull.

3.5.3. Boundary curvature analysis

Once a touch is identified, it remains to draw a line to cut the chromosomes apart. The boundary curvature function (recall *Figure 10*) can be used to identify concavities in the object outline. If two nearby concavities occur close to a node in the skeleton, a cutting line drawn between them will probably separate the chromosomes.

4. Chromosome measurement

Many techniques have been proposed to quantify the salient characteristics of the chromosome image. Generally these are size, centromere position, and features derived from the banding profile. The centromere is usually identified by the associated width constriction. The banding pattern can be described either by global or structural features. Global techniques compute mathematical properties of the banding profile, whereas structural techniques seek to identify and quantify specific visible structures (e.g. bands) in the profile. While both global and structural approaches have met with success, the structural techniques are troubled with problems of identifying the local structures in the first place. Experience indicates that the robustness of global features is an advantage.

4.1 Morphological features

The two most significant features describing chromosome morphology are size and centromere position. These have proved sufficient for dividing the chromosomes into seven classes (groups historically named A–G) but not for identification of the homologous pairs.

4.1.1 Size

Chromosome size is reflected in chromosome *length*, *area*, and *integrated density*. These are the only features that are commonly normalized. They are usually expressed as a percentage of the sum for the entire spread.

4.1.2 Centromere position

The location of the centromere along the chromosome is another valuable piece of morphological information. It is the primary width constriction, and it frequently stains lightly as well, depending on condensation state. The centromere is commonly located as the principal minimum of the width profile. Chromosome *area centromeric index* and chromosome *density centromeric index* are the ratio of chromosome short-arm area (or density) to whole chromosome area (or density).

4.2 Banding pattern features

In order to classify chromosomes one must find ways to quantify the unique characteristics of the band patterns. Certain banding features are orientation-dependent while others are not. That is, they are computed relative to the short and long arms. This, in turn, requires a knowledge of the centromere position to properly orient the chromosome right-side up. Orientation-dependent band features are subject to corruption by centromere location errors. Thus, if its centromere is improperly located, a chromosome is likely to be measured upside down and hence misclassified. Using orientation-independent band features avoids this proliferation of error. For weighting function features (see Section 4.2.3 below), if the weighting function has even symmetry, the feature will be orientation-independent. If the weighting function has odd symmetry, the absolute value of the feature will be orientation-independent.

4.2.1 Fourier coefficients

Shortly after developing Q-banding for human chromosomes (71), Caspersson's group used the low-frequency coefficients of the discrete Fourier transform of

3: Image analysis: quantitative interpretation of chromosome images

the banding profile as features (3). Each of these is actually an inner product between the banding profile and one of the sine or cosine basis functions of the Fourier transform. While this technique was functional, its classification accuracy proved disappointing (3). Fourier classification can approach 90% accuracy on good quality spreads, but these features do not separate the classes well on routine clinical material.

4.2.2 Gaussian distribution functions

Granlund's 'distribution function' approach modelled the banding profile as a sum of shifted Gaussian functions (6, 10, 11, 16). This structural technique fits Gaussian functions to the banding profile and uses the resulting parameters (amplitude, width, position) as features. While mathematically elegant, and intuitively appealing, this approach was found to lack robustness and did not find its way into widespread use.

4.2.3 Weighted density distributions

Granum (72) computed band features as inner products of the banding profile with the set of six heuristically defined 'weighted density distribution' (WDD) functions shown in *Figure 11*. This technique demonstrated good performance on published data sets (24, 32), and it forms the basis for chromosome classification in several commercial karyotyping systems.

The WDD features, developed in (28), are derived by correlating chromosome banding profiles with the set of WDD basis functions. To compute the six WDD features, we first compute the density profile from each chromo-

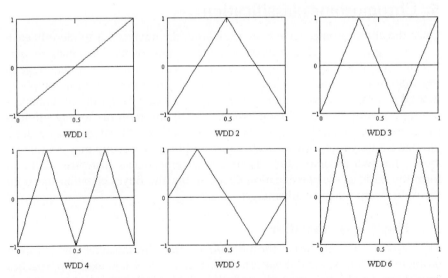

Figure 11. The six weighted density distribution functions that are used to characterize the banding profile.

some by averaging image intensity perpendicular to the chromosome medial axis. If we denote the chromosome density profile as $p(x)$, the WDD features are computed by taking the inner product between $p(x)$ and one of the WDD basis functions $w_n(x)$

$$\mathrm{WDD}_n = \sum_x w_n(x) \cdot p(x) \qquad [9]$$

Since these features are computed as inner products of the banding profile with the WDDs, they can be viewed as coefficients of a linear transformation, where the WDDs play the role of basis functions. This classifier outperforms the Fourier and Granlund classifiers on published data sets, suggesting that the WDDs are better weighting functions for chromosome classification than either sinusoids or Gaussians. Their lack of orthogonality, however, leads to statistical dependence among the features, and they have not been shown to be optimal for chromosome classification.

4.2.4 Heuristic features

Geneticists apparently identify chromosomes using *landmark bands*. Following this approach, one can define features such as 'location of largest band above the centromere', or 'density of darkest band below the centromere', etc. These features perform well, but since they have multimodal probability density functions (pdfs), classifiers that assume Gaussian statistics cannot be used. Normally a non-parametric classifier (one which does not assume a particular mathematical form for the feature statistics) must be used (see Chapter 4).

5. Chromosome classification

Once the chromosomes have been measured, the next step is to identify each one. The process of classification in general, and of chromosomes in particular, is dealt with in detail in Chapter 4. As classification is a critical stage in the analysis of chromosome images, we include an outline here for completeness. We consider classification in two steps: first calculating the probabilities of group membership, and then assigning each chromosome to a position in the karyotype image based on those probabilities. The latter can be done more accurately if one imposes the constraint that each group may contain no more than two chromosomes. There are several logical structures that can form the basis of the classification decision, but the Bayes classifier has both theoretical and practical advantages.

5.1 The Bayes classifier

The most commonly used formulation of the chromosome classification problem is that of statistical pattern recognition (36). While many approaches have been tried, the most robust and reliable is the classical maximum-likelihood classifier based on Bayes' rule.

3: Image analysis: quantitative interpretation of chromosome images

It is common practice to assume that the features have a multivariate normal distribution. While this assumption is difficult to justify experimentally, it usually results in good performance as long as the marginal (single-feature) pdfs are unimodal and reasonably symmetric. The joint pdf under this assumption is given, for class i, by

$$p_i(\mathbf{x}) = -\frac{1}{\sqrt{(2\pi)^M |\mathbf{S}_i|}} \exp[-\tfrac{1}{2}(\mathbf{x} - \boldsymbol{\mu}_i)^T (\mathbf{S}_i^{-1})(\mathbf{x} - \boldsymbol{\mu}_i)] \qquad [10]$$

where $\mathbf{x}$ is the feature vector of length M (the number of features in use), and $\boldsymbol{\mu}_i$ and $\mathbf{S}_i$ are the mean vector and covariance matrix, respectively, for class i.

During the classifier training phase, one first characterizes the classes by computing an M-element mean vector for each class, i, given by

$$\boldsymbol{\mu}_i = [\mu_j]_i = \frac{1}{N_i} \sum_{k=1}^{N_i} x_{i,j,k} \qquad [11]$$

where N_i is the number of class i objects in the training set and $x_{i,j,k}$ is the measurement value for the jth feature of the kth object in class i. One then computes the $M \times M$ covariance matrix for each class. It is given, for class i, by

$$\mathbf{S}_i = [S_{j_1,j_2}]_i = \frac{1}{N_i - 1} \left(\sum_{k=1}^{N_i} x_{i,j_1,k} x_{i,j_2,k} - N_i \mu_{i,j_1} \mu_{i,j_2} \right) \qquad [12]$$

5.1.1 Bayes' rule

If $\mathbf{x}$ is the feature vector of an unknown object, the likelihood that it belongs to class i is given by Bayes' rule

$$L_i(\mathbf{x}) = P_i \frac{p_i(\mathbf{x})}{p(\mathbf{x})} \quad \text{where} \quad p(\mathbf{x}) = \sum_{i=1}^{N_c} p_i(\mathbf{x}) \qquad [13]$$

P_i is the *a priori* (before measurement) probability of occurrence of class i, and N_c is the number of classes. This formula uses what is known about the object (the *a priori* probability, the pdf, and the feature vector) to compute the *a posteriori* (after measurement) probability that the object belongs to class i.

5.1.2 Classification

If we take the log of *Equation 12*, it becomes the 'log likelihood' of class i membership

$$LL_i(\mathbf{x}) = \ln(P_i) - \tfrac{1}{2} |(\mathbf{x} - \boldsymbol{\mu}_i)^T (\mathbf{S}_i^{-1})(\mathbf{x} - \boldsymbol{\mu}_i)| - \frac{M}{2} \ln(2\pi) - \tfrac{1}{2} \ln(|\mathbf{S}_i|) - \ln[p(\mathbf{x})] \quad [14]$$

which can also be used for classification since it is monotonic with likelihood. The third and fifth terms are constant among classes, and can be dropped from consideration. The second term,

$$D_i = \tfrac{1}{2} |(\mathbf{x} - \boldsymbol{\mu}_i)^T (\mathbf{S}_i^{-1})(\mathbf{x} - \boldsymbol{\mu}_i)| \qquad [15]$$

is known as the square of the 'Mahalanobis distance' from the object to the mean of class *i*. It can, by itself, be used for classification if the prior probabilities are either equal or unimportant, and if the fourth term of *Equation 13* takes on similar values among the different classes. A further simplification, one that reduces training set size requirements, is to assume that the S_i matrices are diagonal, that is that the features are uncorrelated within each class. One simply sets the off-diagonal elements of the covariance matrices to zero. The remaining (diagonal) elements of the covariance matrices are simply the feature variances.

5.1.3 Classifier training

If the chromosome features have a distribution that is basically unimodal and reasonably symmetrical, the conditional probability density function for an individual chromosome belonging to class *i* can be approximated by the multivariate Gaussian distribution in *Equation 9*. In general, the mean vectors, μ_i, and the covariance matrices, S_i, must be obtained from a statistical training exercise. Their maximum likelihood estimates can be computed from *Equations 10* and *11*, respectively. These are only estimates, however, of the true underlying values that characterize the chromosome population at large. Each such estimate comes from a binomial distribution which, for a reasonable sample size ($N > 24$), is approximately Gaussian. The RMS estimation error due to statistical sampling error is

$$\sigma_e = \frac{\sigma_s}{\sqrt{N}} \qquad [16]$$

where σ_s is the standard deviation of the population.

Protocol 10. Classification

A. *Design the classifier*

1. Establish the prior probabilities of each class.
2. Select a set of features (measurable properties) that reflect the similarities and differences of objects in the same and different classes, respectively.
3. Eliminate features that exhibit broad intraclass variance, as compared to their interclass variance.
4. Eliminate features that exhibit high (positive or negative) correlation with other features.

B. *Train the classifier*

1. Develop a sufficiently large training set of correctly classified objects.
2. Compute the mean values for the features in each class (*Equation 10*).

3: Image analysis: quantitative interpretation of chromosome images

3. Compute the covariance matrix for the features in each class (*Equation 11*).

C. *Run the classifier*
1. Compute the feature values for each unknown object.
2. Compute the likelihood that each unknown object belongs to each class (e.g. *Equation 13*).
3. Assign the objects to their most likely classes, subject to the prevailing constraints (e.g. only two chromosomes of each type).

6. Karyotype generation

Geneticists use the ISCN (International Standard for Chromosome Nomenclature) ideograms to compare their patient's chromosomes against the normal human chromosome banding pattern. Automated karyotyping systems facilitate this by straightening, orienting, and enhancing the chromosome images and displaying them in karyotype format, optionally beside a properly scaled ISCN ideogram.

6.1 Chromosome straightening

Once the medial axis (skeleton) has been determined for a properly segmented chromosome, it can be straightened by a geometric operation (36). This technique generates a set of axis normals, lines perpendicular to the medial axis, spaced one pixel apart along the chromosome axis. The greylevels along those axis normals are copied into a rectangular array, producing the image of a straight chromosome.

The set of equally spaced points along the axis, as well as the points on each axis normal, generally do not fall upon integer pixel coordinates. Thus it is necessary to use an interpolation process to determine the greylevel at these intermediate points. Bilinear interpolation, the two-dimensional generalization of linear interpolation, generally works well, given the oversampling common in metaphase images. If the slight smoothing effect of bilinear interpolation is objectionable, higher order interpolation (36) can be used.

6.2 Chromosome enhancement

In cytogenetics it is important to be able to visualize subtle details of the banding pattern. Indeed, it is the resolution limit of the light microscope that places the fundamental limit on the sensitivity of this technique. A digitized image, displayed on a CRT monitor, usually offers less detail than that seen through the eyepieces. Without image enhancement, currently available cytogenetics automation instruments would be generally inferior to photomicroscopy in their ability to support the detection of band pattern alterations.

With image enhancement, done prior to the display or printing of the final karyotype image, users normally report equivalent or even improved detection of chromosomal variation when compared with photomicroscopy. The resulting improvement in displayed chromosome image quality, particularly in the detail of the banding pattern, significantly increases the utility and cost-effectiveness of these systems.

6.2.1 Highpass filtering

Two-dimensional convolution filtering can be employed to increase the visibility of subtle image detail (36). Several commercially available automated systems for assisting the preparation of karyotype images make use of convolution filtering to enhance the detail of the banding patterns of individual chromosomes (5, 18–20). Due to the resolution limit of the light microscope, typical chromosome images cannot contain information at frequencies above about 4 cycles per μm. The banding detail of interest occurs a mid-range frequencies (1–2 cycles per μm), and the highest frequencies are usually dominated by noise. Small convolution kernels (e.g. 5×5) can be designed to have transfer functions that start from 1.0 at zero frequency, increase to 5.0 or so at medium frequencies, and fall back towards zero at the folding frequency. The application of such a filter can not only recover the detail lost through digitization and display, but can produce images showing more visible detail than the original. It is convenient to make the amount of mid-frequency boost an adjustable parameter so that the user can tailor the degree of enhancement to the specimen, the application, and personal taste.

6.2.2 Wavelet enhancement

While convolution filtering has long been used with great success for digital image enhancement, considerable interest has evolved recently in the signal processing field regarding new transform techniques that address specific feature detection. These are called (1) multiresolution or multiscale analysis, (2) time-frequency or space-frequency analysis, (3) pyramid algorithms, or (4) wavelet transforms (73).

The Fourier transform uses, as its orthonormal basis functions, sinusoidal *waves*. Each of these basis functions is non-zero over its entire domain. By contrast, many important image features (e.g. chromosome bands) are highly localized in spatial position and are non-zero only over a short interval. Such components do not resemble the Fourier basis functions, and thus are not represented compactly in the transform coefficients (i.e. the frequency spectrum). It is only by intricate constructive and destructive interference that the Fourier transform is able to synthesize these narrow structures with a summation of sinusoids.

To combat this deficiency, mathematicians and engineers have explored several approaches using transforms having basis functions of limited duration that vary in position as well as in frequency. These waves of limited duration

3: Image analysis: quantitative interpretation of chromosome images

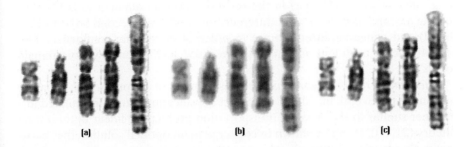

Figure 12. Chromosome image enhancement: (a) high-pass filtered; (b) original; (c) wavelet enhanced.

are referred to as '*wavelets*', and transformations based on them are called '*wavelet transforms*'.

The Harr transform (36, 74) is the earliest example of what we now call a wavelet transform. It differs from the Fourier transform in that its basis functions are all generated by translation and scaling of the Harr function, which is a pair of adjacent positive and negative rectangular pulses. In recent years, new theory developed in the field of multiresolution analysis makes it practical to exploit the inherent advantages of wavelet transforms for image enhancement.

The wavelet transform decomposes an image into components of different size, position, and orientation. As with linear filtering in the Fourier frequency domain, one can alter the amplitude of coefficients in the wavelet transform domain prior to an inverse transform. However, unlike the Fourier transform, the wavelet transform offers localization of image components in both position and scale (size). Such a localization allows us to selectively accentuate interesting components in the image at the expense of undesirable ones (34, 75). *Figure 12* shows a chromosome image enhanced by both highpass filtering (a) and wavelet enhancement (c).

6.3 Chromosome assignment

The likelihood of each of 46 chromosomes belonging to each of 24 homologue types (recall Section 5) forms a 46 × 24 matrix. It then remains to assign each chromosome to a particular slot in the karyotype image. Normally one expects exactly two chromosomes of each type, except for X and Y. Enforcing this constraint upon the assignment process significantly reduces the error rate because the largest computed likelihood does not always correspond to the correct chromosome type. There are at least three approaches to constrained chromosome assignment.

Heuristic assignment. The simplest approach is simply to assign each chromosome to its most likely class, while enforcing class membership rules (e.g. only two of each type) in the process. One can augment this by computing a confidence factor (e.g. difference of largest and second largest likelihood) and assigning chromosomes in order of decreasing confidence. This way the least confident assignments are made last, when few slots remain open.

The transportation algorithm. The assignment of chromosomes so as to maximize the overall likelihood of correct classification can be formulated in a manner similar to the classical transportation problem of linear optimization theory (21, 30). It can be shown to converge to an optimal solution that maximizes the overall likelihood, subject to the given constraints. This approach has proven to work well with a distance classifier (32).

Simulated annealing. Simulated annealing is a mathematical technique that models the gradual solidification of molten metal, as it cools, to form a perfect crystal. Under the right conditions, it can be shown to converge eventually upon a global optimum (76, 77). Like the transportation algorithm, this technique can be applied to the constrained chromosome assignment problem. Unlike the transportation algorithm, it can require a very large number of iterations to converge.

7. Other cytogenetics applications

While automatic karyotyping is perhaps the oldest cytogenetics application of digital image analysis, several more recently developed applications have proved valuable as well.

Pinkel *et al.* (78) described chromosome painting using fluorescent in-situ hybridization (FISH). As a result of further development of this technique (79, 80), it is now possible to paint each of the 24 chromosome types with a unique combination of fluorescent labels (81).

Comparative genomic hybridization (CGH) is a sensitive method to identify chromosomal abnormalities (82). It is a FISH technique wherein red-labelled DNA from a specimen of unknown genetics competes with green-labelled DNA from a normal to bind with a normal metaphase.

A fluorescent intensity profile is computed for each chromosome. A profile plot of the red-to-green intensity ratio alongside the chromosome image helps locate areas of deletion or duplication of DNA. Areas containing DNA sequences that are deleted in the subject will shift towards the green, while areas with duplicated sequences will show an increased red-to-green ratio.

The use of whole-chromosome FISH painting probes for metaphase chromosomes has become common in the past few years (83–85). Dual-colour and multicolour painting (86) are commonly used to detect aneuploidies (87) and resolve translocations in metaphase preparations (88, 89). There is increasing interest in multicolour labelling and full-colour karyotyping to determine structural rearrangements more accurately (90).

7.1 Multiplex-FISH

Twenty-four-colour karyotyping (M-FISH) was introduced in 1996 (81, 91). David Ward's group at Yale developed a filter-based approach (92), while Applied Spectral Imaging has brought to market a Spectral Karyotyping ('SKY') system (93). Each chromosome is labelled with a unique combination of five fluors, and each pixel within each chromosome can be classified as to the chromosomal origin of its DNA. There are two approaches.

Colour thresholding. One can use threshold segmentation in the five separate colour channel images to determine which fluors are present, and which are absent, on a pixel-by-pixel basis. Then, since the chromosome labelling scheme is known, a look-up table can be used to identify the chromosomal origin of the DNA.

Colour-space classification. The five intensity values (one from each probe image) form a five-dimensional colour space. Each pixel on each chromosome plots to a point in that space. A five-feature Bayes classifier, for example, can be trained on pixels from manually classified chromosomes from normal specimens. In operation, the classifier assigns each pixel of each chromosome to one of the chromosome classes (1–22, X, Y).

Colour vectors. The intensity of pixels varies significantly within each chromosome, while the ratio of fluor brightnesses remains relatively constant. To combat this, the five-dimensional colour space can be converted to spherical coordinates. The four colour coordinates (angles) turn out to be more discriminating than the intensity (radial) coordinate, and the spherical approach appears to yield better separation of the clusters in feature space than 5D rectangular coordinates.

Colour display. In its simplest form, each classified pixel in each chromosome is assigned a user-defined colour, based on its class. This set of 24 colours is designed to be visually well distinguished. The resulting colour image may be viewed in either its original spread format or as a karyotype. The result is a cartoon-like image in which each chromosome appears as a coloured blob. Alternatively, each classified pixel can be assigned a hue and saturation only, based on its class, but taking its brightness from the DAPI or Alu-banded image. With this display method the chromosomes are visually distinct in colour, yet clearly show the banding pattern as well.

Misclassified pixels show up as coloured specks or blobs inside the chromosomes. Classification accuracy can be improved by assigning all pixels on each axis normal (line perpendicular to the chromosome axis) to the class of the majority on that line. This reduces misclassified pixels without masking information about structural rearrangements, since these typically occur across the width of the chromosome. The karyotype, annotated to show rearrangements, can also be displayed with only the greyscale banding pattern, providing a more conventional-looking picture.

7.2 Metaphase finding

One area where automation can be a labour-saving factor is in the location of metaphase cells on a microscope slide. This is particularly useful where one must study large numbers of slides having few metaphases. Automatic metaphase finding requires an automated microscope with the ability to move the stage, focus automatically, and recognize clusters of small objects. It can be done at lower magnification than automatic karyotyping, typically with 1-μm pixel spacing.

7.2.1 Autofocus

While many methods have been proposed, the most common means to bring a microscope image into focus automatically is to compute a measure of image sharpness at different positions as the stage is moved along the z-axis, selecting the position that maximizes its value. Two of the most useful focus measures are (1) the sum of squared differences of adjacent pixel greylevels, and (2) the integrated brightness above a threshold. A simple hill-climbing algorithm will converge upon the z-axis position of optimum focus. (See also Chapter 2, *Protocol 5*.)

7.2.2 Metaphase recognition

The first step in the image analysis is image segmentation, as with automatic karyotyping. Then nearby objects are associated together to form clusters which may correspond to metaphase spreads. If the centre-to-centre distance between two objects is below some threshold (a few μm), they become associated. Then each cluster is classified as metaphase or non-metaphase, and the metaphases are ranked for quality. Finally, the stage is returned to each metaphase for review by the operator, in order of quality, the best ones being visited first (31).

A Bayes classifier is useful for classifying clusters of objects as metaphase or non-metaphase material. There may be several classes corresponding to metaphases of different quality, and several corresponding to different types of artefact (e.g. stain debris, undivided nuclei, etc.). The features that have proved useful are (1) the number of objects in the cluster, (2) total area, (3) total perimeter, (4) average area, (5) average perimeter, (6) average inter-object distance, (7) total-perimeter-to-total-area ratio, and (8) average perimeter-to-area ratio. The better spreads have larger values of all these parameters than poor quality metaphases. A linear function of the same parameters can be designed to predict metaphase quality for ranking the spreads prior to relocation. As with chromosome classification, the classifier must be trained on an adequate number of manually identified metaphases and artefacts.

8. Image processing software

We conclude this discussion of image analysis for cytogenetics with a brief review of general purpose image processing software packages that are currently available. These can be used for developing and testing algorithms and for routine processing in chromosome analysis or other work. Some of the more popular ones are listed here, roughly in order of increasing cost. In most cases demo versions are available at no charge. The packages differ in their emphasis and in their user interface philosophy, so it is worthwhile to evaluate the software before making a commitment.

8.1 Academic packages

Written originally for the Macintosh®, **NIH Image** is in the public domain and is available as binary or source code, free of charge, from the National Institutes of Health at http://rsb.info.nih.gov/nih-image/. Containing a library of image processing and analysis functions, it has been extended by macros and ported to Microsoft Windows 95®.

A free image processing and analysis program for Windows 95/NT™, **ImageTool** has a built-in scripting language. It is available at http://ddsdx.uthscsa.edu/dig/itdesc.html. It works with Adobe Photoshop plug-ins and Twain scanners. Written in C++, the source code for the executable and a software developers kit are available.

TIM for Windows is a program for manipulating and measuring images. It is available from the Delft Institute of Technology at http://www.ph.tn.tudelft.nl/software.html.

8.2 Commercial packages

LView Pro for Windows is available as shareware at http://www.lview.com. Originally a graphics viewer, the current version (2.1) incorporates a considerable number of image processing functions that are useful for learning and for algorithm development.

SCIL-Image is an image processing system for UNIX/Win95/NT/PowerMac that can be extended with user-written code. A 30-day trial version is available from TNO-TPD at http://www.tpd.tno.nl/TPD/smartsite64.html.

Khoros Pro 2000© is an image processing software development environment originally developed for UNIX machines at the University of New Mexico. Available for Unix and Windows NT from Khoral Research, Inc. (Http://www.khoral.com/), it uses a flowchart-type graphical user interface and has a large, extensible library of image processing functions.

Imaging Research (http://imaging.brocku.ca/) makes the **Analytical Imaging Station**™, a low-cost image analysis software package designed specifically for bioscience image analysis applications on the PC.

Alice™, for scientific and medical image processing, is supplied by Hayden Image Processing Group (http://www.perceptive.com/). It is available for the Macintosh, PowerMac™, Windows 95/NT. Version 3.0 is available for Windows 95/NT.

WiT™, from Logical Vision, Ltd.™ (http://www.logicalvision.com), also has a drag-and-drop data flow GUI and a rather complete, but expandable, library of image processing and analysis functions. It can share the computational load among machines on a network.

Image-Pro Plus™ scientific image analysis software from Media Cybernetics© (http://www.mediacy.com) is sold through authorized dealers. It is available for Windows 95/NT and Macintosh.

Visilog™, from Noesis Vision, Ltd.™ (http://www.cam.org/~noesis) is available for PCs and UNIX machines. It contains a complete library of image processing and analysis functions.

Amerinex Applied Imaging, Inc.™ (http://www.aai.com) produces ***KBVision***™ for UNIX machines and Windows 95™. It has a visual programming environment and a powerful library of processing and analysis functions. It can be purchased in modules to reduce cost. Their ***Aphelion***™ product has extensive algorithm libraries.

References

1. Butler, J. W., Butler, M. K., and Stroud, A. (1964). In *Data acquisition and processing in biology and medicine* (ed. K. Einslein). Pergamon Press, Oxford, 261.
2. Ledley, R. S. (1964). *Science,* **146**, 216.
3. Caspersson, T., Castleman, K. R., Lomakka, G., Modest, E. S., Mollar, A., Nathan, R., Wall, R. J., and Zech, L. (1971). *Experimental Cell Research,* **67**, 233.
4. Bender, M., Kastenbaum, M., Lever, C., and Pelster, D. (1972). *Comput. Biol. Med.,* **2**, 151.
5. Castleman, K. and Wall, R. (1973). In *Nobel Symposium 23—Chromosome Identification* (ed. T. Caspersson). Academic Press, New York, 77.
6. Granlund, G. H. (1973). *J. Theor. Biol.,* **40**, 573.
7. Harris, J., Nasjleti, C., and Kowalski, C. (1973). *Chromosoma,* **40**, 269.
8. Castleman, K. R., Melnyk, J., Frieden, H. J., Persigner, G. W., and Wall, R. J. (1976). *Journal of Reproductive Medicine,* **17**(1), 53.
9. Castleman, K. R., Melynk, J., Frieden, H. J., Persigner, G. W., and Wall, R. J. (1976). *Mutation Research,* **41**, 153.
10. Granlund, G. H. (1976). *IEEE Trans. Biomed. Engr.,* **23**, 183.
11. Granlund, G. H., Zack, G. W., Young, I. T., and Eden, M. (1976). *J. Histochem. Cytochem.,* **24**(1), 160.
12. Lundsteen, C. and Granum, E. (1976). *Ann. Hum. Genet.,* **40**, 87.
13. Oosterlinck, A., van Daele, J., Dom, F., Reynaets, A., and van den Berghe, H. (1977). *J. Histochem. Cytochem.,* **25**, 754.
14. Hazout, S., Venuat, A., Valleron, A., and Rosenfeld, C. (1979). *Human Genetics,* **49**, 133.

15. Leonard, C., Saint-Jean, P., Schoevaert, D., Eydoux, P., Girard, S., and LeGo, R. (1979). *Human Genetics*, **47**, 319.
16. Van den Berg, H. T. C. M., Habbema, J. D. F., de France, H. F., Bakker, H. K., and de Vries, G. A. (1979). *Comput. Biol. Med.*, **9**, 11.
17. Ledley, R., Ing, P., and Lubs, H. (1980). *Comput. Biol. Med.*, **10**, 209.
18. Bruschi, C., Tedeschi, F., Puglishi, P. P., and Marmiroli, N. (1981). *Cytogenet. Cell Genet.*, **29**, 1.
19. Piper, J., Nickolls, P., McLaren, W., Rutovitz, D., Chisholm, A., and Johnstone, I. (1982). *Signal Processing*, **4**, 361.
20. Piper, J. (1982). *Anal. Quant. Cytol.*, **4**, 223.
21. Tso, M. and Graham, J. (1983). *Pattern Recognition Letters*, **1**, 489.
22. Hazout, S., Mignot, J., Guiguft, M., and Valliron, A. J. (1984). *Comput. Biol. Med.*, **14**(1), 63.
23. Mayall, B., Carrano, A., Moore II, D., Ashworth, L., Bennett, D., and Mendelsohn, M. (1984). *Cytometry*, **5**, 376.
24. Lundsteen, C., Gerdes, T., and Maahr, J. (1986). *Cytometry*, **7**, 1.
25. Zimmerman, S., Johnston, D., Arrighi, F., and Rupp, M. (1986). *Comput. Biol. Med.*, **16**(3), 223.
26. Wu, Q., Suetens, P., and Oosterlinck, A. (1987). *Proc. SPIE*, **767**, 400.
27. Lundsteen, C. and Piper, J. (ed.) (1989). *Automation of cytogenetics*. Springer, New York.
28. Piper, J. and Granum, E. (1989). *Cytometry*, **10**, 242.
29. Granum, E. (1980). *Pattern recognition aspect of chromosome analysis*. PhD Thesis, Technical University of Denmark.
30. Tso, M., Kleinschmidt, P., Mitterreitter, I., and Graham, J. (1991). *Pattern Recognition Letters*, **12**, 117.
31. Castleman, K. R. (1992). *J Radiation Res.* (Japan), **33** (Supplement), 124.
32. Kleinschmidt, P., Mitterreitter, I., and Piper, J. (1994). *ZOR – Mathematical Methods of Operations Research*, **40**, 305.
33. Kleinschmidt, P., Mitterreiter, I., and Rank, C. (1994). *Pattern Recognition Letters*, **15**, 87.
34. Wu, Q. and Castleman, K. R. (1996). *Proc. SPIE*, **2825**, 796.
35. Rosenfeld, A. (1970). *Journal of the ACM*, **17**, 146.
36. Castleman, K. R. (1996). *Digital image processing*. Prentice-Hall, New Jersey.
37. Prewitt, J. and Mendelsohn, M. (1966). *Annals of the N. Y. Academy of Sciences*, **128**, 1035.
38. Prewitt, J. (1970). In *Picture processing and psychopictorics* (ed. B. Lipkin and A. Rosenfeld). Academic Press, New York.
39. Weszka, J. (1978). *Computer Graphics and Image Processing*, **7**, 259.
40. Kirsch, R. A. (1971). *Computers in Biomedical Research*, **4**, 315.
41. Marr, D. and Hildreth, E. (1982). *Proc. R. Soc. London, Ser. B*, **207**, 187.
42. Marr, D. (1982). *Vision*. Freeman, San Francisco.
43. Abdou, I. E. and Pratt, W. K. (1979). *Proc. IEEE*, **67**(5), 753.
44. Roberts, L. G. (1977). In *Computer methods in image analysis* (ed. J. K. Aggarwal, R. O. Duda, and A. Rosenfeld). IEEE Press, New York.
45. Davis, L. S. (1975). *CGIP*, **4**, 248.
46. Canny, J., (1986). *IEEE Trans.*, **PAMI-8**(6), 679.
47. Nevatia, R. (1976). *IEEE Trans. Comp.*, **C-25**, 1170.

48. Lester, J. M., Williams, H. A., Weintraub, B. A., and Brenner, J. F. (1978). *Computers in Biology and Medicine*, **8**, 293.
49. Duda, R. E. and Hart, P. E. (1973). *Pattern classification and scene analysis*. Wiley, New York.
50. Brice, C. R. and Fennema, C. L. (1977). In *Computer methods in image analysis* (ed. J. K. Aggarwal, R. O. Duda, and A. Rosenfeld). IEEE Press, New York.
51. Yakamovsky, Y. and Feldman, J. A. (1977). In *Computer methods in image analysis* (ed. J. K. Aggarwal, R. O. Duda, and A. Rosenfeld). IEEE Press, New York.
52. Horowitz, S. L. and Pavlidis, T. (1977) In *Computer methods in image analysis* (ed. J. K. Aggarwal, R. O. Duda, and A. Rosenfeld). IEEE Press, New York.
53. Zucker, S. (1976). *Computer Graphics and Image Processing*, **5**, 382.
54. Kass, M., Witkin, A., and Terzopoulos, D. (1987). In *Proc of First International Conf on Computer Vision*, 259.
55. Leymarie, F. and Levine, M. (1993). *IEEE Trans PAMI*, **15**(6), 617.
56. Williams, D. J. and Shah, M. (1992). *CVGIP: Image Understanding*, **55**(1), 14.
57. Matheron, G. (1975). *Random sets and integral geometry*. Wiley, New York.
58. Serra, J. (1982). *Image analysis and mathematical morphology*. Academic Press, New York.
59. Serra, J. (1986). *Computer Vision, Graphics, and Image Processing*, **35**(3), 283.
60. Serra, J. (ed.) (1988). *Image analysis and mathematical morphology*, 2nd edition. Academic Press, New York,.
61. Giardina, C. R. and Dougherty, E. R. (1988). *Morphological methods in image and signal processing*. Prentice-Hall, Englewood Cliffs.
62. Meyer, F. and Beucher, S. (1990). *J. Visual Comm. and Image Representation*, **1**(1), 21.
63. Dougherty, E. R. (1992). *An introduction to morphological image processing*. SPIE Press, Bellingham, WA.
64. Haralick, R. M., Sternberg, S. R., and Zhuang, X. (1987). *IEEE Trans. Pattern Anal. Machine Intell.*, **PAMI-9**(4), 532.
65. Maragos, P. (1987). *Optical Engineering*, **26**(7), 623.
66. Pratt, W. K. (1991). *Digital image processing*, 2nd edition. Wiley, New York.
67. Mott-Smith, J. C. (1970). *Picture processing and psychopictorics* (ed. B. S. Lipkin and A. Rosenfeld). Academic Press, New York.
68. Arcelli, C. and Sanniti Di Baja, G. (1978). *IEEE Trans. Systems, Man, and Cybernetics*, **SMC-8**(2), 139.
69. Freeman, H. (1977). In *Computer methods in image analysis* (ed. J. K. Aggarwal, R. O. Duda, and A. Rosenfeld). IEEE Press, New York.
70. Blum, H. (1973). *J. Theor. Biol.*, **38**, 205.
71. Caspersson, T., Zech, L., Johansson, C., and Modest, E. J. (1970). *Chromosoma*, **30**, 215.
72. Granum E. (1980). *Pattern Recognition Aspect of Chromosome Analysis. Computerized and Visual Interpretation of Banded Human Chromosomes*. PhD Thesis, Technical University of Denmark,.
73. Pittner, S., Schneid, J., and Ueberhuber, C. W. (1993). *Wavelet Literature Survey*. Technical University of Vienna.
74. Haar, A. (1955). *Math. Annalen*, **5**, 17.
75. Wu, Q., Schulze, M. A., and Castleman, K. R. (1998). *Proceedings of ISCAS'98*.

3: Image analysis: quantitative interpretation of chromosome images

76. Kirkpatrick, S., Gelatt, C. D., Jr., and Vecchi, M. P. (1983). *Science*, **220** (4598), 671.
77. Otten, R. H. J. M. and van Ginneken, L. P. P. P. (1989). *The annealing algorithm*. Kluwer, Boston.
78. Pinkel, D., Straume, T., and Gray, J. W. (1986). *Proc. Natl Acad. Sci. USA*, **83**, 2934.
79. Nederlof, P. M., van der Flier, S., Wiegant, J., Raap, A. K., Tanke, H. J., Ploem, J. S., and van der Ploeg, M. (1990). *Cytometry*, **11**, 126.
80. Nederlof, P. M., van der Flier, S., Vrolijk, J., Tanke, H. J., and Raap, A. K. (1992). *Cytometry*, **13**, 839.
81. Speicher, M. R., Ballard, S. G., and Ward, D. C. (1996). *Nature Genetics*, **12**, 368.
82. Kallioniemi, A., Kallioniemi, O. P., Sudar, D., Rutovitz, D., Gray, J. W., Waldman, F. N., and Pinkle, D. (1992). *Science*, **258**, 818.
83. Coulton, G. (1995). *Histochem J.*, **27**, 1.
84. Trask, B. J. (1991). *Trends in Genetics*, **7**, 149.
85. Cremer, T., Lichter, P., Borden, J., Ward, D. C., and Manuelidis, L. (1988). *Hum. Genet.*, **80**, 235.
86. Wiegant, J., Wiesmeijer, C. C., Hoovers, J. M. N., Schuuring, E., d'Azzo, A., Vrolijk, J., Tanke, H. J., and Raap, A. K. (1993). *Cytogenet. Cell Genet.*, **63**, 73.
87. Eastmond, D. A., Schuler, M., and Rupa, D. S. (1995). *Mutat. Res.*, **348**, 153.
88. Spikes, A. S., Hegmann, K., Smith, J. L., and Shaffer, L. G. (1995). *Am. J. Med. Genet.*, **57**, 31.
89. Nederlof, P. M., Robinson, D., Abuknesha, R., Wiegant, J., Hopman, A. H. N., Tanke, H. J., and Raap, A. K. (1989). *Cytometry*, **10**, 20.
90. Raap, A. K. and Tanke, H. J. (1996). *Bioimaging*, **4**(2), 39.
91. Schröck, E., du Manoir, S., Veldman, T., Schoell, B., Wienberg, J., Ferguson-Smith, M. A., Ning, Y., Ledbetter, D. H., Bar-Am, I., Soenksen, D., Garini, Y., and Ried, T. (1996). *Science*, **273**, 494.
92. Speicher, M. R., Ballard, S. G., and Ward, D. C. (1996). *Bioimaging*, **4**(2), 52.
93. Garini, Y., Macville, M., du Manior, S., Buckwald, R. A., Levi, M., Katzir, N., Wine, D., Bar-Am, I., Schrock, E., Cabib, D., and Reid, T. (1996). *Bioimaging*, **4**(2), 65.

4

Pattern recognition: classification of chromosomes

JIM GRAHAM

1. Introduction

The topic of pattern recognition is often associated with image analysis. Indeed several textbooks combine the topics (1, 2) and the titles are often used interchangeably. While the two topics intersect, they are distinct. Pattern recognition concerns classification: assigning labels to things on the basis of their properties. This is clearly important in image analysis, and Chapter 3 would not have been complete without some discussion of classification. In this chapter I will deal with classification methods and associated issues, which could be applied more widely (e.g. in signal analysis, speech analysis, medical diagnosis, etc.). However, in keeping with the context of the book it will be illustrated by examples from image analysis. There are many textbooks which deal with the topic of pattern recognition in great depth, from the classic work of Duda and Hart (1), through several worthy contributions (2–5) to the recent comprehensive expositions of Ripley (6) and Bishop (7), which relate the topic to neural networks. The purpose of this chapter is to present a 'user's guide', for which I will draw on all of this substantial body of work. I will make little attempt at mathematical or statistical justification and direct myself to those topics which I believe most 'users' (as distinct from pattern recognition practitioners) would need to be aware of. Where appropriate I will guide readers with an appetite for a greater depth of exposition to the appropriate sources. I have tried to keep mathematical and statistical argument to a minimum, but have found it necessary in defining some basic terms to make use of some vector–matrix notation. Where this has occurred I have tried to explain what it means. I do not think that a reader who is familiar with the basic concepts of probability, mean, and covariance will find their understanding impaired.

As an illustrative example, I will retain the chromosome analysis application of Chapter 3. This turns out to be a very useful example, as it involves assigning objects to 24 classes on the basis of several features covering shape and greylevel properties. As a classification problem, this is more complex than most which are encountered in biomedical image analysis. Also, as noted in Chapter 3, chromosome analysis has a long history in pattern recognition and most classification approaches have been applied to chromosomes at some time or other.

In many biological examples, particularly those associated with medical applications, there may only be two classes of interest: normal and abnormal, say. There are particular analytical approaches which apply in these cases, and I deal with some of these towards the end of the chapter.

The process of classification begins with the measurement of an appropriate set of features corresponding to objects in the image. The objects might be identifiable structures, like chromosomes, or individual pixels. The features might be area, length of perimeter, greatest width, number of indentations on the boundary, average wavelength of transmitted light, etc. There are any number of possibilities. One point to note is that the collection of features used for a particular classification may be measured in a number of different units (square micrometres, millimetres, greylevels, dimensionless counts) and therefore the values can be on dramatically different scales. On the basis of these features the decision is made. The parameters of the decision, defining how objects are to be assigned to classes on the basis of features, are determined by the use of a collection of *training* examples. These may have known classes, assigned by an expert (supervised training), or may be used to define the classes (unsupervised training). Setting the parameters may be as simple as setting thresholds on feature values, or may involve modelling their probability density functions or adjusting the weights in a neural network.

Before embarking on a description of different classification approaches, I will first briefly review the chromosome classification problem. The following sections will deal with the classification process from a probabilistic viewpoint, different approaches to performing the classification, including neural networks and issues of feature selection and training. The later sections of the chapter will deal with classifier validation, and include discussions of certain issues particular to the two-class problem, including receiver operating characteristic (ROC) analysis.

2. The chromosome classification problem

The general problem of banded chromosome analysis has been introduced in Chapter 3. Chromosomes appear as distinct bodies in microscope images in the prophase and metaphase stages of cell division. By appropriate staining they can be made to exhibit a pattern of dark and light bands along their lengths. *Figure 1a* of Chapter 3 shows their typical appearance. Biologically, the 46 chromosomes in a cell belong to 24 groups, labelled 1–22, X, and Y. Each of these groups, in a normal cell, contains two homologous chromosomes, with the exception of groups X and Y. These are the sex chromo-

4: Pattern recognition: classification of chromosomes

somes. In a female there are two X chromosomes and the Y group is empty. In a male the X and Y groups have one chromosome each. Assigning the chromosomes to their appropriate groups is a critical part of the analysis, as the clinical significance of abnormalities, such as structural changes or incorrect numbers of chromosomes, depends on the chromosome groups involved. It is common practice in carrying out the analysis of chromosome appearance (*karyotyping*) to construct a tabular display of chromosomes aligned in their groups (e.g. *Figure 1b* of Chapter 3).

The pattern of bands produced by staining is characteristic of the individual groups and could, in principle, be used to perform the classification. In practice, there is a great deal of variability from cell to cell in the size of the chromosomes and the detailed appearance of the bands. Two other features are useful for classification: the relative sizes of chromosomes within a cell and the position of a characteristic constriction called the centromere.

There are two problems to be solved. First, how the chromosome features are to be represented, particularly the banding pattern. Second, how these features are to be used to assign chromosomes to classes, while coping with the natural variability of chromosome appearance from cell to cell.

The first problem is that of *feature selection*. The important issue about features is that they should provide a representation of the data which retains as much information as possible, and should be stable in the sense that they should not be influenced greatly by the presence of measurement noise. In the case of the banding pattern this is strongly influenced by the linear organization of the bands along the chromosome. The pattern can be expressed succinctly by projecting the chromosome density onto a central (possibly curved) axis, creating a *density profile*. Some examples of these are shown in *Figure 1*.

The second problem is that of classification, and that is the topic of the next section.

(a)

(b)

Figure 1. Density profiles obtained by projecting the banding pattern onto the chromosome centreline for (a) a number 1 chromosome and (b) a number 17. Compare with *Figure 1b* of Chapter 3.

3. Classification methods
3.1 Defining classification in terms of probabilities

It has already been mentioned in Chapter 3 that Bayes' formula (*Equation 1*) provides the probabilistic framework for classification.

$$P(C_i|\mathbf{x}) = \frac{P(C_i)P(\mathbf{x}|C_i)}{P(\mathbf{x})} \qquad [1]$$

In classification we have some feature or set of features, **x**, which we observe for an object. In the notation of *Equation 1*, **x** is a vector, which consists of n elements $(x_1, x_2, \ldots, x_n)$. These elements are the values of whatever features we have selected. We wish to assign the object to one of a set of classes C_i. In the chromosome case the subscript i will take one of the values between 1 and 24. Putting the problem in probabilistic terms, we wish to find the class which has the highest probability given the observed features **x**: $P(C_i|\mathbf{x})$. Bayes' formula tells us how to calculate this probability for any class. The terms on the right-hand side of the equation are:

$P(C_i)$—the prior probability of observing an object which belongs to class C_i, whatever its features,

$P(\mathbf{x}|C_i)$—the probability of finding feature(s) **x** for an object in class C_i (the class-conditional probabilities),

$P(\mathbf{x})$—the probability of finding the feature vector **x** for any object. This is a normalizing constant that ensures that $P(C_i|\mathbf{x})$ sums to unity for all possible values of **x** for all classes, i.e. is a proper probability.

For classification purposes, where we seek to compare the *relative* probabilities of assignment to classes, given a particular observation **x**, $P(\mathbf{x})$ is the same, whatever class the object is being assigned to, and may be ignored in the comparison.

The prior probability is important. In some classification problems, certain classes may occur much less frequently than others, and Bayes' formula provides us with a way of taking that into account. If we consider the priors to be equal, then assignment to classes can be made on the basis of the class-conditional probabilities or *likelihoods*, and we have a maximum likelihood classifier. If the priors are not equal it is important to estimate them, as maximum likelihood classification could lead to serious classification errors. It may be that ready estimates of the prior probabilities exist in the form of knowledge of the problem domain. In the case of chromosome analysis, for example, the prior probabilities for each class in a normal cell are equal, $P(x|C_i) = 1/23$, unless the cell is male in which case $P(X) = P(Y) = 1/46$. In other cases $P(C_i)$ may not be known in advance, and it may be difficult to estimate it for all classes. If there is a large number of classes for example, and some of them have a very low probability of occurrence, it may be impractical to estimate the probabilities. However, it is in just these cases that it is important to have some estimate of $P(C_i)$. Even an educated guess may be better than applying a maximum likelihood decision.

In coming to a classification we need to optimize something. As we have described it, the classification seeks to optimize the *posterior Bayesian probability*, that is the probability of an object belonging to a class as estimated by Bayes' formula. If we simplify the problem by assuming all priors are equal or irrelevant, we attempt to maximize the overall likelihood. In principle, the classifier should be attempting to minimize the *Bayes risk*. This is a quantity which takes into account the cost of misclassifications as well as the probability of class assignment. It can be calculated by assigning a cost to each misclassification (and presumably zero cost to correct classification). In practice, costs can be very difficult to estimate, particularly for multiclass problems. If we assume that all misclassification costs are equal, then maximizing the overall posterior probability achieves the desired result. For a more complete discussion of the Bayes risk, see refs 1 and 6.

3.2 Using Bayes' formula to construct decision surfaces
3.2.1 Box classifier

Consider the simplest classification problem imaginable: two classes C_1 and C_2 (normal and abnormal, say) to be distinguished by a single feature x (size, for example). To apply Bayes' formula, we need to estimate the class-conditional probabilities (likelihoods) of each class and the prior probabilities of observing each class. We can do this by constructing a training set of observations. We find the value x for a number of objects, and ask an expert on the objects to assign each one to its appropriate class. The expert may be using more information about the object than just the measured feature. There may be properties of the objects accessible to the human expert, which our image analysis cannot measure or which for some reason we have chosen not to measure. The important thing is that the assignment is as accurate as possible. (As an aside, it is generally assumed that expert classification is the 'gold standard'. In fact, human experts also make erroneous assignments. Usually these errors are small compared to the errors in automatic classification, as in the case of chromosome analysis. They may not be, however, and it might be appropriate to try to estimate their size by, for example, using a panel of experts and comparing their assignments.)

If we have enough observations we could simply construct a histogram showing the frequency of occurrence of each value of x in each class, and read off $P(x|C_i)$ for any observed x. As we rarely have enough training data to do this it is more common to use the observations to calculate the parameters of some distribution to which we believe the data belong. By far the most common distribution to use for this purpose is the normal distribution, and the parameters we estimate are the mean and standard deviation for the two classes: μ_1, σ_1 and μ_2, σ_2. Thus, from the training data we can estimate $P(x|C_1)$ and $P(x|C_2)$ for any value of x and $P(C_1)$ and $P(C_2)$. Note that the likelihood for class 1 is given by *Equation 2*.

$$P(x|C_1) = \frac{1}{\sqrt{2\pi}\sigma} e^{-\frac{1}{2}\frac{(x+\mu_1)^2}{\sigma_1^2}} \qquad [2]$$

As we have noted, for classification we do not need to have strict probabilities; if we are comparing class likelihoods we can make the computation easier by taking logarithms of *Equation 2*:

$$\log_e P(x|C_1) = -\frac{1}{2}\log_e 2\pi - \log_e \sigma_1 - \frac{1}{2}\frac{(x-\mu_1)^2}{\sigma_1^2} \qquad [3]$$

This does not affect classification, since if $P(x|C_1) > P(x|C_2)$, then $\log_e P(x|C_1) > \log_e P(x|C_2)$. The first term on the right-hand side of *Equation 3* is a constant, and can be ignored for the purpose of distinguishing between classes. The likelihood of class membership is therefore determined by a constant proportional to the standard deviation of the class and the square of the distance of the feature from the class mean. The contribution of the distance to the classification is weighted by the standard deviation of the class distribution. This is intuitively reasonable. To apply the prior probability we need to add $\log_e P(x|C_1)$.

Applying Bayes' formula is equivalent to setting a threshold x_t on x, and assigning an object to C_1 or C_2 according to whether x is greater or less than x_t. *Figure 2* illustrates the point. The class conditional probability distributions overlap. There are certain values of x which occur in both distributions. By setting a threshold we determine whether we will assign these items to one distribution or the other. Bayes' formula allows us to determine a value of x_t which will give the minimum classification error on the training data. The

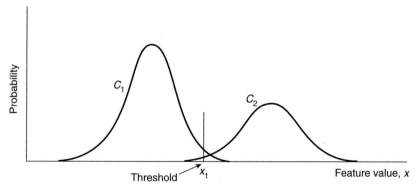

Figure 2. Setting a decision threshold to separate two classes based on a single feature x. The feature distributions are modelled as normal.

threshold illustrated in *Figure 2* will not produce a minimum classification error, but will bias the classification to misclassify fewer C_2 objects at the expense of a higher rate of misclassifications of C_1 objects. This may be appropriate if the costs of misclassification of C_2 objects is greater than the cost of misclassification of C_1 objects.

Extending this approach to two or more features and several classes results in a *box classifier*. *Figure 3* shows that we can construct a space in which each axis corresponds to the possible values of one of the features. (This space is variously referred to as *feature space* or *decision space* or *classification space*.) If we have two features x_1 and x_2 the decision space is two-dimensional. If we have more than two features, the space takes on higher dimensions, but becomes difficult to draw. Here we can simply make a classification decision by comparing each observed feature x_i with its corresponding threshold x_{it} (see *Figure 3*). We have, in effect, constructed a *decision surface* in the form of a box defined by the thresholds on each axis. Box classifiers may be useful in some circumstances; classification is straightforward and quick to calculate. However, it is clear from the illustration on *Figure 3* that they do not in

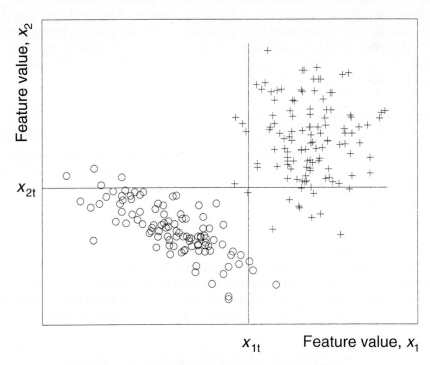

Figure 3. A box classifier. A two-dimensional feature space shows the feature values of two classes, indicated by circles and crosses. Setting thresholds on each axis allows a classification to be made. The decision surface is a box. In this case, a number of misclassifications will occur, wherever the thresholds are placed.

general give a good approximation to maximizing the posterior Bayesian probability.

3.2.2 Linear and quadratic classifiers

Given some training data such as that illustrated in *Figure 3*, Bayes' rule tells us that we wish to estimate $P(\mathbf{x}|C)$, where $\mathbf{x}$ is a vector of two or more features. As in the single-feature case we could either construct a multidimensional histogram from which we read off the probabilities or, more likely, use the training data to estimate the parameters of a model. This model will usually be a multivariate normal distribution. In multivariate statistics, the equivalent of the variance is the covariance matrix, $\mathbf{S}$. If, for example, we were classifying with three features, the covariance matrix is given by *Equation 4*.

$$\mathbf{S} = \begin{bmatrix} \sigma_1^2 & \sigma_{12} & \sigma_{13} \\ \sigma_{12} & \sigma_2^2 & \sigma_{23} \\ \sigma_{13} & \sigma_{23} & \sigma_3^2 \end{bmatrix} \quad [4]$$

σ_1^2 is the variance of the first feature, σ_{12} is the covariance of features 1 and 2, etc. (The matrix is symmetric about the principal diagonal because $\sigma_{12} = \sigma_{21}$.) Effectively, the covariance matrix defines a rotation of the principal axes of the normal distribution away from the feature axes. We can extend *Equation 3* to the multivariate case as shown in *Equation 5*, in which $\mathbf{x}$ and $\boldsymbol{\mu}_i$ are vectors representing the features and the class means respectively. We need to include the dimension of the feature space, d.

$$\log_e P(\mathbf{x}|C_i) = \frac{d}{2}\log_e 2\pi - \frac{1}{2}\log_e |\mathbf{S}_i| - \frac{1}{2}(\mathbf{x} - \boldsymbol{\mu}_1)^T \mathbf{S}_i^{-1}(\mathbf{x} - \boldsymbol{\mu}_i) + \log_e P(C_1) \quad [5]$$

In this equation, the matrix algebra notation may not be familiar to everyone. The third term on the right-hand side is the equivalent of the squared weighted distance function in *Equation 3*. We can write this distance function as in *Equation 6*.

$$D^2(\mathbf{x}, C_i) = (\mathbf{x} - \boldsymbol{\mu}_1)^T \mathbf{S}^{-1}(\mathbf{x} - \boldsymbol{\mu}_i) \quad [6]$$

The distance $D(\mathbf{x}, C_i)$ is called the *Mahalanobis distance*. We can make the notation a little more intuitive by considering the case where all the features are uncorrelated with each other (i.e. the axes of the distribution are parallel to the feature axes). In this case the off-diagonal elements of the covariance matrix are zero and, for the three-feature example, *Equation 6* would reduce to *Equation 7*.

$$D^2(\mathbf{x}, C_i) = \frac{(x - \mu_1)^2}{\sigma_1^2} + \frac{(x_2 - \mu_2)^2}{\sigma_2^2} + \frac{(x_3 - \mu_3)^2}{\sigma_3^2} \quad [7]$$

Viewed in this way, *Equation 7* is reasonably intuitive. The distance of a feature from the mean of a class distribution is just the Euclidean distance

4: Pattern recognition: classification of chromosomes

scaled by the variance along each dimension. *Equation 6* calculates the same property, taking the correlations into account.

We can ignore the constant term in *Equation 5*, just as we did for *Equation 3*, and construct a *discriminant* function (*Equation 8*) that will allow us to distinguish between classes.

$$g_i(\mathbf{x}) = -\frac{1}{2}(\mathbf{x} - \boldsymbol{\mu}_i)^T \mathbf{S}_i^{-1} (\mathbf{x} - \boldsymbol{\mu}_i) - \frac{1}{2}\log_e |\mathbf{S}_i| + \log_e P(C_i) \qquad [8]$$

It is possible to make this discrimination function simpler by assuming that the covariance matrix is the same for all classes. Using this 'pooled' co-variance matrix, the $\log_e|\mathbf{S}|$ term becomes constant for all classes, and can also be dropped (*Equation 9*).

$$g_i(\mathbf{x}) = -\frac{1}{2}(\mathbf{x} - \boldsymbol{\mu}_i)^T \mathbf{S}^{-1} (\mathbf{x} - \boldsymbol{\mu}_i) + \log_e P(C_i) \qquad [9]$$

We can use either *Equation 8* or *Equation 9* for class discrimination. It turns out that, because the covariances are pooled, the discrimination function of *Equation 9* is equivalent to drawing a linear decision surface between the classes (*Figure 4*). In two dimensions (two-feature classifier) this is a straight line, in three dimensions it is a plane, and for higher dimensions it is a hyperplane. For this reason it is called a *linear classifier* or *linear discriminant*. If *Equation 8* is used, the decision surface is quadratic and we have a *quadratic classifier*.

Pooling the covariance matrix may seem a perverse thing to do. Since the individual class distributions are modelled in the quadratic classifier, it ought to give us more accurate classifications. The reason it may be useful to pool the covariances concerns the process of forming the inverse of the covariance matrix in calculating the Mahalanobis distance. If the number of training examples in each class is small, it may not be possible to calculate the inverse matrix $\mathbf{S}_i^{-1}$ for each class. By pooling the covariance matrix, we construct it from more examples, which enables us to invert it and construct a classifier from the training data at the expense of a possible loss of classification accuracy. Another, more technical, reason for pooling the covariances arises from the source of variability. If the variation in the observed feature values arises from an intrinsic property of the data, then using class-specific covariances is most appropriate, provided there are enough training data for numerical stability. If, on the other hand, the variation arises from a basic imprecision in the measuring method, which would apply to objects of all classes, then using a single covariance matrix for all classes is indeed the most appropriate thing to do. As usual, in practice it can be difficult to separate these different cases.

Before concluding I should stress that the terms linear and quadratic classifier are used here to refer to a particular form of model used to parameterize the distribution, namely the multivariate gaussian (normal) distribution.

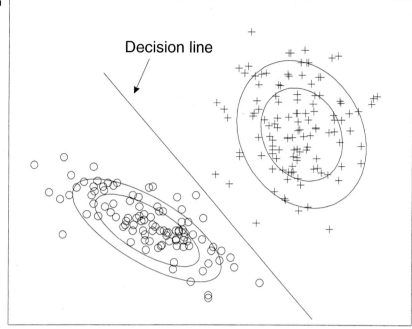

Figure 4. A linear classifier. If the training sets of the two classes are modelled as multivariate normal distributions, the classification can be made using *Equation 9*, which is equivalent to generating a decision surface in the form of a straight line. In this two-dimensional case the distributions are bivariate and the ellipses indicate contours of the normal distributions.

Protocol 1. Classification using a linear or quadratic classifier

A. *Train the classifier*
- Collect feature measurements from as large a set as possible of objects to be classified.
- Have an expert assign each object to a class as accurately as possible.
- Estimate the prior probabilities of each class, $P(C_i)$. This may be done by simply counting the occurrences in the training set (if the set has been sampled without bias—a 'stratified random sample') or using some domain knowledge.
- For each class, i, calculate the mean, μ_i, and covariance matrix S_i. Ensure that the inverse of the covariance matrix (S_i^{-1}) can be calculated for each class. If it can, calculate $\tfrac{1}{2}\log_e |S_i|$ for each class and use the

4: Pattern recognition: classification of chromosomes

class covariances in the following steps (quadratic classifier), otherwise pool the measurements to generate a covariance matrix **S** (linear classifier).

B. *Classify a new (unseen) object*
- Measure the feature vector, **x**.
- Calculate the Mahalanobis distance $D^2(\mathbf{x}, C_i)$ to all class means, using either the pooled or class covariance matrices.
- Linear classifier: if the prior probabilities are equal, assign **x** to the class with the nearest mean. If the priors are not equal, assign **x** to the class with the highest discriminant function obtained using *Equation 9*.
- Quadratic classifier: calculate the discriminant function of *Equation 8* for each class, omitting the last term if the priors are equal.

3.3 Non-parametric methods
3.3.1 Estimating the probability distributions

The linear and quadratic classifiers described in the previous section are *parametric* classifiers. They are characterized by the parameters of the assumed underlying distributions. This may not always be useful. For example, there may not be sufficient training data to make reliable estimates of the parameters, or it may be known that the class likelihood distributions do not conform to a readily parameterized model. In such a case we could turn to non-parametric methods. For example, if, as was suggested, we used the histograms themselves to obtain likelihood estimates, we would have a non-parametric Bayesian classifier. This would require the training data to cover the feature space rather densely. If there were not sufficient data, we could artificially fill the training space by using each training datum as the centre of a local distribution (possibly a normal distribution, but simpler local distributions can be used, such as rectangular or triangular distributions). These local distributions would overlap, allowing the likelihood values to be read from the distribution at places where there were no training examples. These are the so-called *kernel methods*.

Alternatively, we could model the training data, not as a single normal distribution, but as a several overlapping normal distributions. (Kernel methods are the limiting case of this.) The expectation–maximization (EM) algorithm (8) provides a method of doing this, and provides a powerful way of estimating complex distributions. However, its description is beyond the scope of this chapter. A more detailed account of it can be found in ref. 9.

These methods are non-parametric in the sense that the likelihoods are not defined by the parameters of the underlying distribution. In using them, however, we are not freed from parameters. Even if we simply use the data histogram, we need to decide on appropriate bin sizes. In the case of the kernel methods, we would have to specify the form and width of the kernels, and whether that width should be constant, or vary in some way with the data. (Ideally, this should be estimated from an assessment of the measurement accuracy.) The EM algorithm requires us to specify how many normal distributions are to be fitted to the data. These parameters are all rather arbitrary and considerable experimentation may be required to arrive at suitable values.

3.3.2 Nearest neighbour classification

A commonly used form of non-parametric classifier is the *nearest neighbour classifier*. The idea is extremely straightforward. We have a training set of expertly classified data as usual, but we do not attempt to characterize the distribution in any way. In this case it is often called the *reference set*. For any new object that we wish to classify, we simply calculate the distance in feature space between the object's feature vector and each feature vector in the reference set. The new object is assigned to the class of the object in the reference set for which the distance is smallest. As shown in *Figure 5*, the decision surface for a nearest neighbour classifier is piecewise linear.

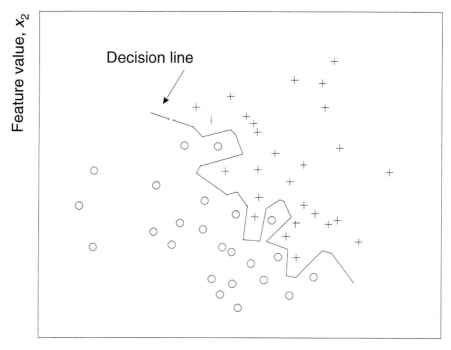

Figure 5. A nearest neighbour classifier. The decision surface bisects neighbouring training items from the two classes, and is piecewise linear.

4: Pattern recognition: classification of chromosomes

As the classification result could be critically dependent on the values of individual items in the reference set, it is potentially sensitive to outliers among the training data. This is usually dealt with by selecting a number of nearest neighbours in the reference set, which then 'vote' for the classification. This gives us the *k-nearest neighbour* (or *k*-NN) classifier, where *k* is the number of vectors in the reference set that are used for a classification decision. As with other non-parametric methods, there is a parameter to be set, namely the value of *k*. Part of the reason for using the *k*-NN classifier is that the distributions may have odd shapes, so that it makes most sense to use the parts of the distribution local to the place where the decision is being made. This argues that *k* should not be too large. If the decision on a particular classification is split, it is likely to be between two classes, so it makes sense for *k* to be odd to avoid ties. Values of 3, 5, 7, and 9 for *k* are typical. As usual, the most appropriate value will be determined empirically.

Another matter which must be decided is the distance metric to be used. The obvious one to use is the Euclidean distance (*Equation* 10).

$$D^2(\mathbf{x},\mathbf{y}) = (\mathbf{x}-\mathbf{y})^T (\mathbf{x}-\mathbf{y}) = (x_1-y_1)^2 + (x_2-y_2)^2 + (x_3-y_3)^2 + \cdots (x_n-y_n)^2 \quad [10]$$

Here the vector to be classified is **x** and the reference set consists of a set of vectors $\mathbf{y}_i$ ($i = 1, \ldots n$, where there are *n* members of the reference set). There is a problem with the Euclidean metric, in that each of the different features are likely to be measured on different scales (they are 'non-commensurate'— see Section 1). If one of the features, for example, happens to be expressed in large numbers, it will dominate the distance measure. It is therefore necessary to use normalized features for classification. There are several normalization schemes, which could be applied, but a sensible approach would be to rescale all the features to have a standard deviation of 1.0. In fact, the really sensible thing to do is to use the Mahalanobis distance as the distance metric. It needs to be borne in mind, however, that this distance has to be calculated to every vector in the reference set from every vector to be classified. In some practical situations, computation time may be important, and scaling the individual standard deviations may be a reasonable compromise (equivalent to assuming all the off-diagonal elements of the covariance matrix are zero).

If time is really critical, another distance metric might be used, such as the sum of the absolute values of the differences in the individual feature values (the 'city block' distance). In this case it might be more sensible to normalize the values according to the mean absolute distance from their median (2).

There are many possible modifications of the *k*-NN approach, such as weighting the votes of the reference vector according to their distance from the test vector. Also, since the decision surface is determined by the reference examples near the class boundaries (see *Figure 5*), execution times may be reduced by selectively removing some of the reference examples. Several of these are explored in refs 2 and 10.

It can be demonstrated that the *k*-NN classifier is suboptimal in comparison to the Bayesian classifier, but that its error rate approaches that of the Bayesian classifier as both k and the number of samples in the reference set become large (1, 6). This may be helpful to know in some circumstances, but the k-NN classifier is often used precisely because a large training set is not available.

Protocol 2. *k*-Nearest neighbour classification

A. *Train the classifier*
- Collect feature measurements from as large a set as possible of objects to be classified.
- Have an expert assign each object to a class as accurately as possible.
- Retain the entire set of training data as a reference set.
- Decide on a distance metric. If using the scaled Euclidean distance (see text), calculate the variance of each feature (σ^2).
- Decide on a value of k.

B. *Classify a new (unseen) object*
- Measure the feature vector, **x**.
- Calculate the distance to each vector in the reference set ($D^2(\mathbf{x}, \mathbf{y}_i)$). Perform the scaling (e.g. $D^2(\mathbf{x}, \mathbf{y}_i)/\sigma^2$ for scaled Euclidean distance).
- Select the k examples in the reference set for which the scaled distance metric is smallest.
- Assign **x** to the class which is most represented among these. In case of a tie, assign to the class with the closest example.

---

4. Features and feature selection

The usefulness of a classifier is to a large extent dependent on the features, which are chosen. Of all the properties of an object that we could measure, we wish to find the set which gives the best discrimination between classes. As Ripley (6) points out, 'this is an impossible problem; there may be no substitute for trying them all and seeing how well the resulting classifier works'. This is not an encouraging statement. For even a modest number of features and training examples, the task of trying all possible combinations is computationally out of the question. Fortunately, we often know something of the problem and the type of features that human experts use in carrying out classifications.

The difficulty may be in translating a feature that is visually accessible to a

4: Pattern recognition: classification of chromosomes

human being into measurements which can be made on an image. This is particularly true of shape descriptions. Many elementary image analysis texts and commercial systems provide ready measurements of shape, such as circularity (square of perimeter length divided by area), aspect ratio (length of the longest chord divided by the length of the longest chord perpendicular to it), etc. Such measures tend to be rather rough and ready, and it may be necessary to apply some ingenuity in image processing methods to obtain sensible representations. Binary morphological operators (Chapter 3) are often useful here—subtraction of a thresholded image from the result of performing a closing operation can reveal indentations round the boundary, for example. As mentioned earlier, this chapter is not concerned with the processing of images. It is sufficient to note that care needs to be taken in which features are measured and how that measurement is done. The key requirements are stability and reproducibility.

One approach is to measure everything one can think of and hope the classifier will deal with it. This is unlikely to be a good strategy. As we will see later, there is a limit to the number of features that can usefully be applied in practice before classification performance deteriorates. This is particularly the case if features are used which do not themselves give good discrimination between classes. There is also the danger of selecting highly correlated features. For example, in some sources we may find the measure of circularity defined above. A circular object will give a value of 4π for this measure, while the more non-circular the object is, the higher the value will be. In other sources (maybe even the same ones) we might find a measure of 'compactness' defined as area divided by the square of the perimeter length. Individually, these are sensible enough ways of trying to represent intuitive descriptions of shape. One is, however, the inverse of the other, and they are therefore completely correlated. It would be foolish to use both as features in the same classification. In fact, the correlation is non-linear, and may be rather difficult for feature extraction methods (Section 4.2) to identify.

An informal examination of the features being measured should be made to ensure that the measurements make a reasonable representation of the required properties and that none of the measurements are inappropriate. In the case of chromosomes, there are two morphological features used by cytogeneticists in addition to the banding pattern: the size and the centromere position. Size could be measured either as length or area. These are different measures, but highly correlated, so that it is sensible to use only one. The centromere divides the chromosome into two 'arms', one of which is usually shorter than the other. The centromere position can be represented as the ratio of the size of the shorter arm to the size of the whole chromosome (the centromeric index). These sizes again could be measured in terms of length or area. Since the short arms may be roughly equidimensional, it may be difficult to find an image analysis method which would measure length reliably, so basing the centromeric index on area measurements may be most useful.

4.1 Selecting features

Having informally selected a set of features we may wish to reduce their number by finding a subset which achieves the same, or nearly the same, classification performance. Indeed, if we have insufficient training data, using too many features may impair classification performance (see Section 6.1). Several texts deal with methods for achieving this (see, for example, refs 3, 11, and 12). These methods often become impractical when there are more than a few features to be evaluated. With larger feature sets *stepwise selection* methods may be used, of which there are three variants. The first of these is *forward stepwise feature selection*. In this technique we take our features one at a time, and use them to perform a classification in isolation. We select the feature on the basis of some discrimination measure (the selection function). The criterion for inclusion in the feature set is that a new feature should improve the value of the selection function by some predetermined amount. The first feature selected is the one which gives the largest value of the selection function when used in isolation. We then go through all the remaining features to evaluate the selection function when used in combination with the first in a two-feature classifier, selecting the one which produces the best value of the selection function, and so on. We continue adding one feature at a time until we include all the features, or no remaining feature meets the inclusion criterion. The disadvantage of this method is that there is no consideration of inter-relationships with variables not yet selected, and once a variable is selected, it cannot be removed.

In *backward stepwise selection* we do the reverse—start with all features, and one by one remove from the list those whose removal causes the least decrease in the selection function. The criterion for feature removal is that the selection function should be reduced by no more than a specified amount. This approach has the advantage over forward selection that the performance of the retained features is judged in relation to the complete feature set. It is, however, computationally more expensive and once a variable has been removed it cannot be reselected.

The third variant seeks to overcome some of the disadvantages of the other two by combining them. It is called (perhaps unsurprisingly) *forward and backward stepwise selection*, and is outlined in *Protocol 3*.

Protocol 3. Forward and backward stepwise feature selection

1. All features are initially excluded from the selection function.
2. The first feature included is the one with the largest value of the selection function.
3. All other features are re-evaluated for inclusion and the feature giving

4: Pattern recognition: classification of chromosomes

the best value of the selection function in combination with the first is retained.

4. The feature entered first is evaluated for removal, and removed if it meets the removal criterion.
5. All features not in the model are evaluated for inclusion, and the one producing the largest value in combination with those currently in the model is included.
6. All features in the model are evaluated for removal. Those meeting the removal criterion are removed.
7. Repeat steps 5 and 6 until no more features meet the inclusion or removal criteria.

One matter to be decided is the nature of the selection function. There are several possible functions which seek to measure the separation between class distributions. The simplest may be the square of the Mahalanobis distance between the closest two classes. One widely used measure is *Wilks' lambda*, which seeks to combine between-class separation and within-class cohesiveness (see ref. 12). It is defined in *Equation 11*.

$$\lambda = \frac{|\mathbf{W}|}{|\mathbf{B}+\mathbf{W}|} = \frac{\sum_{i=1}^{g}\sum_{j=1}^{n_j}(\mathbf{x}_{ij} - \boldsymbol{\mu}_i)(\mathbf{x}_{ij} - \boldsymbol{\mu}_i)^T}{\sum_{i=1}^{g}\sum_{j=1}^{n_j}(\mathbf{x}_{ij} - \boldsymbol{\mu})(\mathbf{x}_{ij} - \boldsymbol{\mu})^T} \quad [11]$$

$\mathbf{B}$ is the between-groups covariance matrix, $\mathbf{W}$ is the within-groups covariance matrix, $\mathbf{x}_{ij}$ is the jth example in class i, $\boldsymbol{\mu}_i$ is the mean vector of class i, $\boldsymbol{\mu}$ is the mean of all classes, g is the number of classes, and n_i is the number of examples in class i. A large between-group covariance (separability) and a small within-group covariance (cohesiveness) result in a large value of λ.

None of the stepwise methods will produce an optimal set of features, unless we are lucky. To do that, we need to exhaustively explore all possibilities, which is impossible. A potentially useful alternative approach is to use one of the more recently developed search techniques designed to find good, if not necessarily optimal, solutions in large search spaces. One of these is *genetic search*. In this technique, the dimensions of the space which are being searched (in this case the features) are encoded in linear arrangements called chromosomes. (This nomenclature is a little unfortunate in the current context.) The chromosomes replicate and are subjected to transformations analogous to those occurring in cell division (such as crossover and point mutation), and the new chromosomes are evaluated for fitness, in this case by the value of a function, such as Wilks' lambda or their classification performance. The 'fittest' chromosomes proceed to the next generation. After a number of generations only the fittest chromosomes survive: those representing combinations of features which give good classification. Genetic search is a mechanism for exploring large spaces which allows the thorough exploration of interesting regions of the space, but provides a mechanism for discovering other interesting regions to explore (13, 14). Another approach to this is *simulated annealing*, where the analogy is that of finding a minimum energy state as an alloy cools. Both of these methods require a careful definition of what is to be optimized. They have advantages and disadvantages compared to each other, depending on the data set being analysed. It is beyond the scope of this chapter to discuss them further, but the interested reader may consult refs 9 and 15.

4.2 Combining features

It may be that, due to unavoidable correlations in the features, the dimension of the decision space could be reduced by forming linear combinations of features. If some of the features are highly correlated, the observed variation in the data may be described in a space of lower dimension. One way to achieve this is to apply *principal component analysis* (PCA). The method is described in Appendix B of Chapter 7 and will not be described in detail here. In mathematical terms, it involves a standard linear transformation, namely determining the *eigenvectors*, of the covariance matrix. The result is to find the direction in feature space of the vector along which the variance in the data is largest, the vector perpendicular to that along which the variance is next largest, and so on. This produces a number of vectors, which is the same as the original number of dimensions, but translated and rotated to coincide with the distribution of features (see *Figure 16* of Chapter 7). Since these vectors are ordered in terms of the quantity of variance they account for, we can reduce the dimensionality by ignoring those vectors that account for only a very small proportion of the total variance. Thus we can finish up using all of the original measurements, but in combination so that the number of classification features is smaller. One advantage of the technique is that there will be no linear correlations among the resulting features. Non-linear correlations will not be eliminated however, so it would not help in identifying our circularity/compactness blunder. The method is sometimes known as *eigenanalysis*, and sometimes appears as the *Karhunen–Loève expansion* in the pattern recognition or communication theory literature.

An important point to stress about PCA is that it conducts an analysis of the variance of the data. If some of this variance is due to measurement error (noise), that error will be reflected in the answers. There are also problems with non-commensurate data. A related technique is *factor analysis*, which seeks to generate a representation of lower dimension by examining the correlations among features. The main advantage over PCA is that factor analysis deals explicitly with the noise in the data. However, factor analysis requires non-linear optimization to arrive at a solution, while PCA is solved straightforwardly using linear mathematics (16).

4: Pattern recognition: classification of chromosomes

Principal component analysis is a linear method of projecting data from a space of high dimension to one of lower dimension. *Sammon mapping* (17) tries to find a non-linear mapping to a space of lower dimension in such a way that the intersample distances are maintained as far as possible in the lower space. It does this by minimizing an error function consisting of a sum of the differences between the intersample distances in each space. This has also been used as a basis for classification (e.g. ref. 18), and is described in ref. 4.

4.3 Clustering

Clustering is closely related to feature combination. In this case we are not necessarily interested in reducing the dimensionality of the data. Rather we wish to identify whether the data form natural groups in feature space, thus organizing themselves into classes, which may not have been identified beforehand. We can consider this as a form of unsupervised training, where we seek clusters of similar data, which we can label as classes to which new feature vectors can be assigned, by any of the means already described.

There are many clustering methods, which are described in refs 1–5. Here I describe two commonly used methods, *k*-means and its extension ISODATA. The *k*-means algorithm appears in a number of different flavours, the most common of which (19) is shown in *Protocol 4*.

Protocol 4. The *k*-means algorithm

1. Select a number of clusters required (*k*).
2. Allocate *k* points in feature space as cluster centres, for example by seeding the clusters on the first *k* data items or applying some domain knowledge.
3. For each of the remaining items find the closest centre and assign the item to that cluster.
4. Recalculate the cluster means (centroids) from the vectors assigned to each cluster.
5. Repeat steps 3 and 4 for all the data, moving items between clusters as necessary.
6. Stop when there is no further movement between clusters.

In the algorithm shown in *Protocol 4*, the cluster centroids are recomputed simultaneously, once all the data items have been assigned to clusters. If the training data set is large, several iterations may be required for the algorithm to converge. In a variant of the algorithm (20), described in ref. 2, the cluster centroids get recomputed as soon as a new item is added. One further pass is then required to reassign the items, without updating the cluster centres. The algorithm in *Protocol 4* will produce a globally optimum assignment, subject to the number of classes. The modification produces a local optimum.

A disadvantage of the k-means algorithm is that the number of clusters must be specified beforehand. It may be possible to estimate this number by inspection or by some knowledge of the problem. In fact it is possible for clusters to finish up with no members, so the final number of clusters may be less than k.

The ISODATA algorithm (21) seeks to address this problem by applying heuristic rules for discarding, splitting, and merging clusters, so that the final result provides a number of clusters within some range around the desired number k. The rules are governed by a number of parameters, which must be specified beforehand. As with the parameter k in the k-means algorithm, these may need to be set empirically. The algorithm is described in *Protocol 5*.

Protocol 5. The ISODATA algorithm

Define values for the parameters:

N_c the desired number of clusters,

N_{min} the minimum number of samples allowed in a cluster,

D_{min} the minimum distance allowed between cluster centroids,

S_{max} a parameter which controls splitting, usually some fraction to be applied to the standard deviation of each feature value,

M_{max} the maximum number of merges allowed per iteration.

Assign the items to k clusters using the k-means algorithm. Then iterate as follows:

1. Discard clusters with fewer than N_{min} items. Discard their data items also. Reduce k accordingly.
2. If $k > 2 \times N_c$ or the iteration is even, try to merge clusters.
3. If $k < N_c/2$ or the iteration is odd, try to split clusters.
4. Recalculate the cluster centroids and repeat from step 1 until no changes occur in the clusters or some maximum number of iterations is reached.

Cluster splitting

Find a cluster which has a standard deviation in some feature x greater than $S_{max} \times \sigma_x$, where σ_x is the standard deviation of the feature over the whole data set. Separate this cluster into two sets, those with x greater than the mean and those with x less than the mean. Calculate the centroids of these clusters. Increase k.

Cluster merging

If the distance between two centroids is less than D_{min}, merge the clusters. Calculate the centroid of the merged cluster and reduce k. Repeat for all clusters to be merged up to M_{max} times.

Steps 2 and 3 of the algorithm seek to force merging, if the number of clusters is more than twice the desired number, or splitting, if the number is less than half the desired number. Of course, these limits could be changed by applying some further heuristic parameters.

5. Neural networks

5.1 Introduction to neural networks

From a pattern recognition point of view, we cannot ignore neural networks; indeed some of the most comprehensive recent books on pattern recognition specifically relate the properties of classical statistical methods to neural networks. On the other hand, the topic is very large, and there are texts dealing with the topic at every level, from the introductory to the advanced (6, 7, 22–25). Here, even more than in previous sections, I assume that the reader does not intend to build a neural network *ab initio*. There are several tools around for doing that (see Section 8.1). Rather, I take the reader to be a user and I will attempt a functional description of neural networks from a pattern recognition point of view without being greatly concerned to describe the inner mechanics further than is necessary to understand their properties.

The name neural networks is used to cover a wide variety of processing architectures which involve simple processing units amongst which there is a large number of connections by weighted links. Only a few of these even take inspiration from models of processing in true neurons, and we might more accurately talk of parallel distributed processors. However, neural networks is a snappier name, and it has become current. Practitioners in the topic try to cover the embarrassment by speaking of *artificial neural networks* (ANNs).

The general form of neural networks of interest for pattern recognition is shown in *Figure 6*. A number of nodes (or units) act as the input layer, each receiving the value of one of the features. The input nodes distribute the feature values along links to the next layer, which may be the output layer. Each input node is connected to each node in the second layer. The links between nodes have associated weights. Thus, the input to each node in the second layer is a vector consisting of all the input values, each multiplied by an element of a weight vector. Each node in the second layer has a different weight vector associated with it, corresponding to the weights on its links. The second layer nodes calculate some transformation of the weighted inputs, each producing a single output. This vector of output values may be the output of the network, or these outputs may be passed on to another layer in the

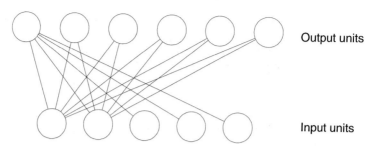

Figure 6. The arrangement of nodes and links in a feedforward neural network. Each node in the input layer is connected to each node in the next layer (which may be the output layer). For clarity only some links are shown.

same way, the second layer being fully connected to the third, which may generate the overall output or pass its outputs on to yet another layer. The number of nodes in the input layer is determined by the number of features. If we are using the network for classification, the number of nodes in the output layer may be the same as, or greater than, the number of classes. There may be any number of nodes in the intervening layers (known as *hidden layers*).

Networks with the simple layered organization described here are called *feedforward* networks. There are forms of network with more complex connectivity, but this form is common to the networks most widely used for classification. The nature of the network is determined by the transformation which takes place at the nodes. A few different examples are described in the following sections. Irrespective of the network, this transformation is the same at all nodes (except the input nodes). The properties of the network that make it specific for a particular task are determined by the pattern of weights. *Training* the network consists of setting the weight values appropriately using a training set of feature vectors. Two styles of training are available: supervised and unsupervised training, which are used for different purposes. In supervised training, the classes of the training data are known and used to set weights so that the correct class of output is generated for a new vector. This is similar to training a parametric classifier. In unsupervised training, the classes are not known, and the purpose of the training is to discover clusters of the training data in feature space. These clusters can be labelled, either by defining them as classes or attaching class labels to them after training. New vectors are then classified according to the clusters they most resemble.

5.2 Supervised training

5.2.1 The multilayer perceptron

The *multilayer perceptron* (MLP) has a structure in which one or more hidden layers intervene between the input and output layers. Each node applies a non-linear transformation to the sum of its (weighted) inputs. Each output node corresponds to a class and the values of the outputs correspond to likelihoods of the input vector belonging to each class. After successful training, presentation of a new set of inputs should produce the highest output value at the node corresponding to the correct class. Training occurs by presenting each feature vector in the training set as input, calculating the difference between the observed output vector and the desired output vector (specified by the correct class), and using this difference, or error function, to alter the weights on the links to the output nodes. The error function is propagated back through the layers, adjusting each layer of weights in turn. The training data are presented to the network many times (in different *epochs* of training). Each time the configuration of weights moves closer to a stable configuration in which the overall net error is a minimum. Training is controlled by two parameters: the *gain* or *learning rate*, which determines the amount by which the weights change in each backpropagation pass, and the *momentum* which stabilizes the convergence by preventing the weight changes on one pass being very different from the weight changes on the previous pass. These parameters affect the rate of convergence, but also the final weight configuration and hence classification performance. The nature of the training algorithm is such that it can have several undesirable properties. It tends to be slow as the weight values change incrementally with each training epoch. It may converge to a local minimum in the overall net error (i.e. find a weight configuration which produces a net error lower than similar weight configurations, but is not the lowest that could be found). It may 'oscillate' (successively find weight configurations which are close to the minimum error configuration, without ever reaching it). There are several variants of the basic training algorithm which attempt to address one or other of these problems, but all involve backpropagation of error and training in epochs.

After training, the network weights should be configured so that on presentation of a new vector, not included in the training set, the output node corresponding to the correct class should have a higher value than all the others. Ideally, this node should output the value 1, while the others output 0, but as with other classifiers, this is unlikely. Indeed, it is common practice to train networks with a desired output of, say, 0.9 and 0.1, rather than 1.0 and 0.0.

Protocol 6.   Multilayer perceptron training

Collect two annotated sets of data: a training set and a test set.

Select sensible values for the gain and momentum parameters (e.g. 0.1, 0.5).

Set the number of input nodes to match the number of features.

Set the number of output nodes to equal the number of classes.

Select a number of nodes in the hidden layer.

Protocol 6. *Continued*

Present all data to the network. At each presentation readjust the weights using an appropriate backpropagation algorithm.

Monitor the net error.

Keep training until the net error stops decreasing.

Experiment with reducing gain. For example, if the net error is not reducing significantly, halve the gain value and continue training. Also evaluate the network by classifying the test data. Halve the gain if the total number of classification errors ceases to decrease. Stop when the gain has been reduced to some tiny value.

Experiment with different values of momentum and initial gain.

Experiment with different numbers of nodes in the hidden layer.

Experiment with more than one hidden layer. (More than one is probably unnecessary, and you certainly do not need more than two.)

Use the test data to evaluate different network and parameter configurations. (See the note in Section 7.4 about classifier validation.)

5.2.2 Radial basis function networks

It is not necessarily helpful to think of *radial basis functions* (RBFs) in terms of a network at all. However, they are often presented that way, and sometimes considered to be an alternative to the MLP. The RBFs are a collection of functions, each of which has a centre, and produces a result which depends only on the distance from the centre. A good example of this is the gaussian function (normal distribution) with equal values of σ in all directions, but there are many other radially symmetric functions that could be used. If we distribute the centres of these functions around the feature space, then any new vector, which corresponds to a point in feature space, will produce a function value according to its position relative to each of them. If we associate each centre with a class (perhaps more than one with each class), and multiply the function value by a suitable weight, then the result can be arranged to be highest for the correct class. RBF classifiers form a variant of the kernel methods described in section 3.3.1.

In network terms, if we associate each RBF with a single output node, then the weights on the links provide the weighting values for the RBF outputs. Training can be done by a least-squares method (using matrix inversion, rather than iteration) and takes place in a single pass.

There are some parameters to be set: the positions of the centres in feature space and the scale factor. The centres may be distributed uniformly over the feature space, but this will not result in an efficient use of them, particularly if the dimension of the space is large. They may be placed in regions of the space where training vectors occur using a clustering algorithm, such as *k*-

means, or placed according to the positions of classes. The scale factor (σ if the RBFs are gaussian, or the equivalent values for other functions) is generally taken to be the same for all centres. Note that as we are considering distances, we are again dealing with the scaling issue we encountered in the *k*-NN classifier.

5.3 Unsupervised training
5.3.1 Competitive networks
Competitive networks have the form shown in *Figure 6*. In this case there is no transformation of the output, rather the outputs are compared with each other. It has been noted that the set of weights can be considered to be a vector associated with each output node. The number of elements in the weight vector is, of course, the same as the number of elements in the feature vector. That is to say the dimension of the weight vector is the same as the dimension of the feature space, and the weight vector can be considered to be a point in feature space. Furthermore, the element by element multiplication of feature values by weight values followed by summation of the products defines the inner (or dot) product of two vectors **w·x**. The dot product can be used as a measure of the distance between two vectors (or points). So summing the weighted feature values results in a number corresponding to the distance between the feature vector and the weight vector in feature space.

The operation of the competitive network is to try to alter the positions of the weight vectors to move closer to the feature vectors. Training proceeds by presenting each feature vector in turn. At each presentation, each of the output nodes contains a value corresponding to the distance between the node's weight vector and the feature vector. The node for which the distance is smallest is chosen as the winner. The weights corresponding to that node only are altered. (This training algorithm is sometimes called a *winner-take-all* algorithm.) The weight values are altered in such a way as to move the weight vector closer to the feature vector. *Figure 7* illustrates this for weight vectors and data vectors in a 2D space. As the feature vectors are successively presented, different nodes win the competition, and their vectors shift into positions which coincide with clusters in the feature vectors. This says precisely what the competitive network does. It identifies clusters in the training data. We could subsequently label the nodes with the classes appropriate to the clusters, or interpret the clusters as classes which have been discovered in the training data by the competitive algorithm. Of course, some nodes will never win the competition, or do so only rarely. These can be pruned from the network following training.

In classification, a new vector on being presented to the network will finish up closer to one of the trained weight vectors than any other, and be classified according to the label assigned to the associated node.

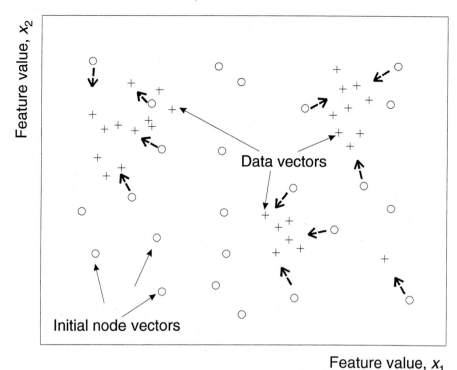

Figure 7. Competitive network training. The distribution of weight vectors on the input nodes and training data vectors in feature space are represented by circles and crosses respectively. The node vectors are initially random, but shift their positions as shown by the broken arrows as they respond to the winner-take-all training algorithm.

5.3.2 Kohonen's self-organizing map

A widely known version of the competitive network is Kohonen's *self-organizing map*. In this version, the output nodes can be considered as belonging to a two-dimensional array (*Figure 8*), in which each node in the output layer (the competitive nodes) is aware of its neighbourhood, i.e. its immediately adjacent nodes, and their neighbours out to some distance. The winner-take-all algorithm proceeds as described in the previous section, with the additional step that nodes in the neighbourhood of the winning node are also moved. In this way, the nodes that identify the clusters tend to form physically adjacent groups on the map. As training proceeds, the size of each node's neighbourhood and the amount by which the vectors are moved are reduced.

The result is a map in which regions correspond to clusters in the input data, or discovered classes, which can be labelled.

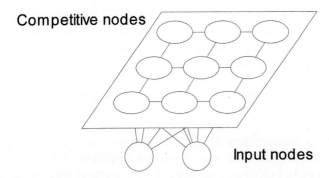

Figure 8. The arrangement of nodes in the Kohonen self-organizing map. As in other feedforward arrangements, the input nodes are fully connected to the output nodes. The nodes in the output layer are arranged to have a geometric relationship to each other, so that each node 'knows' its neighbours out to some specified distance.

6. Classifying chromosomes

Here I consider how some of the methods described in the preceding sections have been used in classifying chromosomes. I will review the features which have been used to describe chromosomes and different forms of classifier—maximum likelihood classifiers and MLP neural networks. As a case study it is revealing, since it demonstrates certain practical issues concerning classifier training and evaluation.

6.1 Features

I have already discussed the nature of some of the features available for chromosomes. The morphological features of size and centromeric index can be used to perform a partial classification into seven groups. The banding features, which are necessary to assign a chromosome to one of the 24 biologically significant groups, are based around the density profile. In the long history of work on chromosome classification, several representations of the banding pattern have been proposed. One early proposal (26) was to fit a series of gaussian functions to the profile peaks, and use the parameters of the gaussians (μ and σ) as classification features. Groen et al. (27) suggested another scheme which used the shape, density, and positions of specific bands on the image, rather than the density profile.

The most successful band pattern descriptors have been used in the most recent studies. The weighted density distributions (WDDs) introduced by Granum (28) and refined by Piper and Granum (29) involve multiplying the density profile by a series of functions of the form shown in *Figure 11* of Chapter 3. The transformed density values corresponding to each of the WDDs, summed along the profile, provide the classification features for a maximum likelihood classifier. In effect, the WDDs form a basis set for the density profiles in the same way as sinusoidal functions for a basis set in Fourier analysis. In this case, the basis functions are defined intuitively to respond to the observed properties of the density profile.

The other recently used band descriptor is simply to use the profile sample values themselves as features. These have been used in neural net classifiers (30–33) and in a quadratic classifier (34). There are, of course, too many profile samples to be useful (well over 100 for some of the longer chromosomes), so the number needs to be reduced. Errington and Graham (30) reduced the number to 15 without loss of accuracy, simply by decimating the profile samples. Nivall (34) used the same approach. Lerner *et al.* (35) used a form of backward feature selection and principal component analysis to choose a subset of banding features, and also arrived at 15 features. Very similar classification performance was obtained whichever feature selection method was used.

In trying to determine the optimum number of WDD features to use, Piper (36) observed the occurrence of a well-known phenomenon known as *peaking*. This is the effect whereby as successively more features are used, classification performance improves (as expected) until a peak is reached, after which the addition of more features results in a reduction in classification accuracy. This initially surprising result arises from the use of a finite (usually small) training set. Adding features increases the dimension of the feature space. The training data find themselves spread out in spaces of higher and higher dimension. Eventually a point is reached at which the training data are no longer adequate to define the decision surfaces properly.

6.2 Data sets

Several data sets of annotated chromosome measurements have been collected and used in classification studies. Each data set contains data on chromosome size, centromeric index, and a sampled banding profile, along with an expertly allocated classification. These data sets can therefore be used for training and as ground truth in evaluating classifiers (see Section 7). They are summarized in *Table 1*. The Copenhagen, Edinburgh, and Philadelphia data sets all contain data from chromosomes at metaphase, where nominally 450 bands are visible in a cell. Owing to different types of material and different

Table 1. Chromosome data sets used in classification experiments

	Copenhagen	Edinburgh	Philadelphia	600-band	CPR
Number of chromosomes	8106	5584	5947	6177	128990
Source tissue	Blood	Blood	Chorionic villus	Blood	Amniotic fluid
Visual quality	High	Routine	Low	Medium	Routine

4: Pattern recognition: classification of chromosomes

Table 2. Summary of results of chromosome classifiers

Classifier	Ref.	Features	Classification performance			
			Copenhagen	Edinburgh	Philadelphia	600-band
Quadratic	37	WDD	5.3	16.9	22.4	8.7
Quadratic	34	Profile samples	–	–	–	8.2
Quadratic	38	Profile samples	5.5	17.3	24.2	7.6
MLP NN	30	Profile samples	5.8	17.0	22.5	–
MLP NN	35	Profile samples	–	16.4	–	–
Constrained	37	WDD	2.0	11.2	14.7	2.8

data acquisition procedures, the visual 'quality' of the chromosome images, and hence of the measurements, varies. The 600 band data were collected from chromosomes at an earlier state of contraction and show more bands. The CPR data set was also collected in Copenhagen from short metaphase chromosomes, as part of the routine use of an automated karyotyping system. The main feature of this data set is its large size.

6.3 Classifiers

Both maximum likelihood and neural network classifiers have been applied to chromosome classification. While there have been a number of such studies, the ones which have achieved the most useful results and which are comparable for the purposes of this chapter are summarized in *Table 2*. The first five rows of the table show 'context-free' classification. Of the many studies using WDD features, the best classification rates are reported in ref. 37. In the studies of Nivall (34), Charters (39), and Errington and Graham (30), a reduced set of band profile features was selected from the density profiles by decimation. Lerner *et al.* (35) used feature selection followed by PCA for feature selection. They also applied their classifier to another data set, which has not been included here, as it cannot be compared directly with the other studies. The last row shows the effect of applying the karyotyping constraint (see section 6.3.2).

The main conclusion that we can draw from *Table 2* is that the all these classifiers achieve rather similar performance. Some of them do better on some data sets than others. The differences are sometimes even significant in a statistical sense, but they are small and not particularly useful from a practical point of view.

6.3.1 Data-set size and the covariance matrix

In some studies (36) using the WDD features, it was observed that the off-diagonal elements of the covariance matrix could be set to zero without significant loss of classification accuracy. This is equivalent to assuming the features are uncorrelated, which is demonstrably not the case. In fact, it is possible to optimize classifier performance by weighting the off-diagonal elements by a factor between 0 and 1 (40). It was later shown that the usefulness of this heuristic arises because of an insufficiency of training data. If the number of training examples is sufficient, it is always better to use the full covariance matrix. However, expertly annotated training data is a precious resource, and the number of examples required to make the training set big enough could be rather large (see Section 7.5). In many circumstances, we have to make do with what we have got, and covariance weighting may be a useful heuristic trick.

6.3.2 The role of context

A particular feature of chromosome classification is that it is constrained. In a normal cell, each class should be assigned exactly two chromosomes (except the X and Y classes in a male cell, which should have one each). This constraint (the 'karyotyping constraint') can be used to improve classification performance by rearranging the assignments following a 'context-free' classification such as those reported in *Table 2*. There have been several approaches to this, but the most successful makes use of an algorithm from operations research, known as the *transportation algorithm*. This exploits the fact that the constraints are linear (i.e. only involve one chromosome and one class each). The transportation algorithm addresses the general problem of distributing goods from a set of sources to a set of destinations along paths which have known costs. The solution that minimizes the overall cost is found by linear programming. In the chromosome context, the sources are the classes, the destinations are the chromosomes, and the costs are the discrimination function values (41). It turns out that there is a particularly efficient solution in the chromosome case, where the destination requires delivery of only one item (42). The transportation algorithm guarantees that the rearrangement among classes achieves the optimum *a posteriori* assignment subject to the constraints. That is not to say that all classifications are correct. The rearrangement is still limited by errors in the context-free classification.

6.3.3 Ultimate classification

In a study which 'pulled out all the stops', Kleinschmidt *et al.* (37) drove down the classification error for classifying chromosomes in the main data sets to values shown in the last row of *Table 2*. To achieve this, they weighted the off-diagonal elements in the covariance matrix by 0.8, used the logarithm of the Mahalanobis distance (rather than negative log-likelihoods) as the classification measure and applied transportation rearrangement. The context-free classification performance is shown in the first row of *Table 2*, and is better than that achieved by other classifiers, although not hugely so. The use of the Mahalanobis distance seems to have suited the transportation algorithm, and the constrained classification rates are much lower than have been achieved in any other study.

7. Classifier validation

7.1 The need for validation

Presumably, we wish to use our classifier on some data that have not been classified by a human expert. It will clearly be helpful to know how well it will do, that is to have some measure of the expected error rate, when applied to unseen data. Maybe we wish to compare different classifiers applied to the same task (as in *Table 2*). We need to present some data to the classifier, and compare its performance with some 'ground truth'. This requires us to have some examples classified by an expert.

We could get an estimate of the error rate by presenting the training data to the classifier, and counting the errors. The error rate we observe by such a 'resubstitution' scheme is called the *apparent error rate*. Clearly, it will give a biased answer in the sense that the measured error rate will be lower than that achieved on unseen data. (In the extreme case of a nearest neighbour classifier, the apparent error rate would always be zero.) We really need a different, independent data set, which has also been classified by an expert to use as a validation set. This can be achieved by dividing the training data into two equal parts, using one for training and the other for classification. In this way, we obtain an *unbiased* estimate of error rate. The estimate may be unbiased, but it may not be very accurate. We are likely to get more accurate estimates if the training set is larger. We could address this by swapping the roles of the training set and the test set, conducting two evaluation experiments. The result will be two unbiased estimates, the average of which will be a better estimate of the true error rate than either. This method is known as *cross-validation*. All of the chromosome classifiers shown in *Table 2* were evaluated in this way.

7.2 Cross-validation, the jackknife and the bootstrap

Using the precious resource of our training data, we can construct better trained classifiers by extending the cross-validation method. If we use, say, three-quarters of the data to train the classifier, and validate the remaining quarter, repeating the procedure till all of the training data has been used for validation, each estimate will be unbiased, since no single item in the data is used for both training and validation. We would get better estimates still by randomly selecting a small validation set and carrying out the cross-validation many times. In the limiting case, we have *leave-one-out* cross-validation, summarized in *Protocol 7*. It is expensive in computational terms as, in principle, a new classifier needs to be trained for every example, although for some classifiers, such as the linear discriminant, this may not be necessary (6). If our classifier is an MLP neural network, however, we could be forgiven for losing our nerve. According to Ripley (6), leave-one-out cross validation is also known as *ordinary* cross-validation.

> **Protocol 7.** Leave-one-out cross-validation
>
> 1. Select an item from the training set. This may be an individual item or a collection of items from all classes, such as all the chromosomes in a given cell.
> 2. Using all the remaining training vectors construct a classifier (e.g. *Protocols 1* or *2* or neural network training).
> 3. Present the selected vector(s) to the trained classifier and classify according to the appropriate rule (nearest neighbour, closest mean, highest network output, etc.).
> 4. Assign an error value to each assignment. This will clearly be zero for a correct assignment. If all misclassifications are considered equally undesirable, a value of 1 can be assigned to each misclassification, otherwise they may be weighted with appropriate cost functions.
> 5. Accumulate the errors in an error sum.
> 6. Repeat steps 1–5 for all items in the training set.
> 7. Divide the error sum by the number of items to produce an average error.

Leave-one-out cross validation is sometimes erroneously referred to as the *jackknife* method. The jackknife in this context is a method for estimating the bias (the overoptimism) in the apparent error rate, which provides a different way of looking at the problem. The confusion between the two methods arises from the fact that in both cases estimates are made by successively missing out items from the training set. In the jackknife case, however, the omitted item is not used as a classification object. Instead, given n items in the training set, n different training sets, each of size $n - 1$ (one item omitted), are used to provide different estimates of the apparent error rate from which the bias can be estimated.

The *bootstrap* (43–45) is an alternative to the jackknife for estimating bias. Instead of leaving a single element out of the training set, a large number of *bootstrap* samples are used, each obtained by random sampling (with replacement) of the full set of training data. Again the bootstrap sets are used to generate different estimates of the apparent error rate, and hence an estimate of the bias. The bootstrap is a promising method, which has not yet been fully evaluated in the pattern recognition literature (but see, for example, ref. 46). Its principal disadvantage is its high computational cost.

7.3 The confusion matrix

The comparison of classifiers in *Table 2* simply used the total error rate as the measure of classification performance. This is reasonable enough, as we generally want a classifier that makes few errors. However, it may be that we are interested in particular types of error, such as confusion between chromosomes X and 7 or 4 and 5. We can obtain such information by constructing a *confusion matrix*. If we have m classes, the confusion matrix M has dimension $m \times m$, and $M(i,j)$ expresses the number of times an item in class i is assigned to class j. The elements may be counts of misclassifications, or normalized to probabilities. It is clearly desirable that the off-diagonal elements of M be as small as possible. Unexpectedly large entries off-diagonal, as part of an otherwise low overall error rate, would indicate that particular types of errors are being made consistently.

7.4 Validation issues for neural networks

In the case of neural networks there is an additional requirement for validation. The weights are set by minimization of some error function with respect to a training set; a separate test set is used to compare networks (e.g. with different numbers of hidden nodes, or trained with different training parameters—see *Protocol 6*). The resulting classifier is therefore defined by both sets, and using the second set as a validation set would result in a biased validation. It is necessary to have a third data set for validation (7).

7.5 Training-set size and validation

How big a training set do we need? The existence of the very large CPR dataset of chromosome measurements (*Table 1*) enabled Piper to carry out a study investigating the effect of training set size on the estimated error rate (47). His premise was the following. If the training set is not large enough, the 'apparent' error rate is clearly optimistically biased. The error rate from a cross-validation experiment, while unbiased with respect to the particular trained classifier, will be pessimistically biased compared to a classifier trained with a much larger set of data. If the training set size increases, the optimistic bias of the apparent error rate should decrease as the training set becomes more representative of the underlying population. Similarly, the pessimistic bias of the cross-validation estimate should decrease. In the limit, where the training set is 'large enough', the two error rates should converge. By conducting an experiment with even larger training sets (and evaluation sets, always the same size as the training set) derived from the CPR data set, he demonstrated that this was indeed the case. From his results he proposed the rule of thumb that the training set size should be about ten times the number of *free parameters* to be set. In the case of a parametric classifier, these are the class means and covariances; in the case of a neural net, the number of weights. This rule of thumb gives a figure for the training set size much larger than is often quoted. One suggestion for the size of training set has been ten times the number of classes. For the chromosome problem, this would mean that 240 training examples would be enough. Piper's proposal is that the number should be about 10000. Errington (48) conducted a similar study on the same data using an MLP neural net classifier and obtained similar results.

It turned out, in Piper's study, that the mean of the apparent and cross-validation error rates was always a good estimate of the asymptotic error rate, i.e. the optimistic and pessimistic biases were always about equal. I know of no way of showing that this is a general result. There are few example applications with such a huge body of annotated data, so it is unlikely that the experiment will be repeated. Errington's study with the MLP classifier was again in broad agreement, although the symmetry between the optimistic and pessimistic classification rates was less marked.

7.6 Validation issues for the two-class problem

7.6.1 Detection

The two-class problem is not different in principle from the many-class problem, but certain specific issues arise. The two-class problem is often posed in terms of identifying normal and abnormal cases, and the context may be one of *detection*, rather than classification. Thought of in that way, certain terminology becomes appropriate.

The confusion matrix becomes a 2 × 2 matrix (see *Figure 9*), the off-diagonal elements being *false positive* (FP) and *false negative* (FN) errors, the diagonals being true positive (TP) and true negative (TN) elements. We need to be a little careful, as what constitutes a positive or negative observation rather depends on what we are seeking to detect. Generally speaking, a positive observation is the detection of an 'abnormal' case. A false positive is a normal case misclassified as abnormal. A false negative is an abnormal case misclassified as normal. In a given study, it may be that false positives or false negatives are more important, and evaluating a classifier solely on the basis of overall misclassification rates is misleading. We can use TP, FP, TN, and FN values to provide measures of detection as follows.

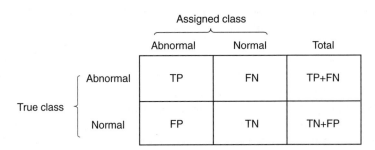

Figure 9. The confusion matrix for a two-class problem (normal versus abnormal). In evaluation of a detector, a fixed number of true normals and true abnormals will be presented (third column). The detector can be evaluated by the proportion of off-diagonal elements in the confusion matrix.

The *true positive fraction*, $\frac{TP}{(TP + FN)}$, gives a measure of how likely it is that a genuinely abnormal item will be detected.

The *true negative fraction*, $\frac{TN}{(TN + FP)}$, gives a measure of the likelihood of a normal item being correctly identified.

The *false positive fraction* is given by $\frac{FP}{(TN + FP)}$.

The *false negative fraction* is given by $\frac{FN}{(TP + FN)}$.

These different measures are clearly inter-related, and it is common to define a detector in terms of just two of them: the true positive fraction and the false positive fraction (sometimes called the *hit rate* and *false alarm rate* respectively). In signal detection theory the *sensitivity* of a detector is defined by combining these two measures. There are several ways of doing this. *Equation 12* defines one such measure, known as d', which uses the z-scores derived from the normal distribution.

$$d' = z(\text{true positive fraction}) - z(\text{false positive fraction}) \qquad [12]$$

A classifier which assigned classes at random would achieve a hit rate and a false alarm rate both of 0.5. As $z(0.5) = 0$, this classifier will give $d' = 0$. Reasonably successful classifiers give values of d' around 1.0, and a perfect classifier (hit rate 1.0 and false alarm rate 0.0) would give an infinite value for d'. Various tricks have been proposed for avoiding this numerical embarrassment (49).

The term *sensitivity* is also used in medical decision making as a synonym for the true positive fraction, expressed as a percentage (50). This is because it measures the proportion of (say) cases in which a disease is present correctly identified by a diagnostic test. The true negative fraction is sometimes called the *specificity*, being the proportion of disease-free cases that are correctly identified. The confusion in terminology is regrettable. Here I use 'sensitivity' in the signal detection sense.

7.6.2 ROC analysis

Setting thresholds for the two-class problem was dealt with in *Figure 2*. In this context, the two classes are normals and abnormals. Unless we are extremely fortunate, there will be overlap in the distributions of the two classes and we will always have some FP and FN errors. Setting the threshold depends on the perceived importance of these two different types of errors. If it is possible to assign relative costs, c_n and c_p, to the two types of error, then it can be shown that the threshold should be set where FP/FN $= c_n/c_p$. However, in practice, costs of misclassification are often very difficult to estimate.

Another method of assessing detection performance, which assumes nothing about relative costs, is the analysis of the *receiver operating characteristic* or ROC. Typical ROC curves are shown in *Figure 10* in which the true positive fraction is plotted against the false positive fraction. Each curve corresponds to a different classifier. As the classification threshold (or *operating point*) is altered, the proportion of different types of errors changes (see *Figure 2*), providing the different points on the curve. Note that provided the underlying assumptions about normality of data are valid, all points on the same curve correspond to the same values of sensitivity (d'). For this reason they are sometimes called *iso-sensitivity curves*. The different proportions of false positive and false negative errors correspond to variations in the *bias* of the classifier. In altering the thresholds, we do not make our classifier more or less sensitive, but we alter its readiness to classify a feature vector as normal or abnormal. Different classifiers will produce different shapes of curve. A classifier that assigned items at random would generate a diagonal line (*Figure 10a*). As it should always be possible to achieve a hit rate of zero by classifying everything as normal, the point

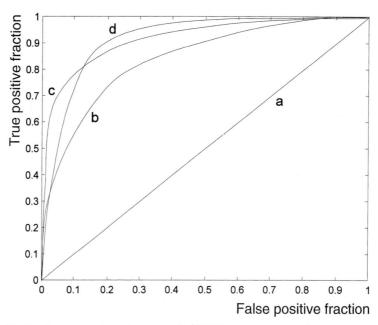

Figure 10. Receiver operating characteristic (ROC) curves. (a) A diagonal line corresponds to a random classifier, which takes no account of the feature values. (b) The more sensitive the detector, the more the curve moves to the top left of the diagram, still passing through (0,0) and (1,1). (c,d) More sensitive detectors still. The classification here may not correspond to the assumptions of normality. The detectors are about equally sensitive, and the choice of detector would be determined by the desired operating point. Detector (c) would be preferred if the priority was to have a low false positive fraction.

(0.0,0.0) should always lie on the curve. Similarly, one could always achieve a perfect hit rate by classifying everything as abnormal, in which case a false alarm rate of 1.0 will also be achieved resulting in the point (1.0,1.0) lying on the curve. (ROC curves which pass through the points (0.0,0.0) and (1.0,1.0) are called *regular*.) More sensitive classifiers have ROC curves which are shifted towards the upper left corner of the space (*Figure 10b–d*). The ROC curve of a perfectly sensitive classifier would pass through the point (0.0,1.0). When ROC curves are calculated empirically, the assumptions of normality are likely to be violated, and the resulting curve may deviate from the ideal. *Figure 10c,d* shows two different classifiers of about equal sensitivity with different shapes of ROC curves.

There are several measures of sensitivity that could be derived from the ROC curve to compare classifiers (49). A commonly used measure is the area under the curve (AUC), which has the value 0.5 for a random classifier and 1.0 for a perfectly sensitive one. A higher value of AUC is clearly desirable. Care should be taken, however, as AUC would not be able to distinguish between the classifiers represented by the ROC curves of *Figures 10c* and *10d*. To select between these, we would need to be clear whether we wished to minimize false negative or false positive errors. In either case, we would have to choose an operating point (classifier threshold), but the ROC curves would provide a quantitative way of making the choice.

In image analysis we often wish to evaluate whether a detection algorithm is capable of identifying particular image characteristics, such as abnormalities on a radiograph. We could validate our detector by applying it to images, in which an expert has annotated the suspect regions. In this case, a detection is only considered successful if the region of image identified as positive by the detector coincides with a region identified by the expert. It is therefore necessary to have some criterion for spatial coincidence—e.g. the 'true' and 'detected' regions share a single pixel, or overlap by 50% of their areas, or are exactly coincident. Once a coincidence criterion has been established, the true positive fraction can be measured. However, every other pixel in all evaluation images is a potential false positive. Measurement of the false positive fraction is difficult since there is a huge number of potential negative classifications. In practice, classifiers can be compared by using a similar curve in which the abscissa is obtained by counting the number of false positives per image. This is called a *free-response operating characteristic* (FROC) curve (free response since the classifier is not given a fixed number of normal objects). The ROC space is no longer square and simple measures like AUC are not very helpful. However, the curve can be useful when comparing two classifiers applied to the same data or the same classifier applied to different subclasses of the data. At a given operating point the better classifier would be placed further to the top left of the space. *Figure 11* shows empirical FROC curves. They were derived from data obtained in a study evaluating methods of detection of lesions in mammograms (51). The two curves represent the

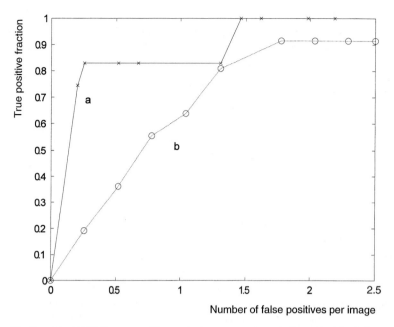

Figure 11. Empirical FROC curves. Although the curves do not show the ideal shape, it is clear that curve (a) corresponds to better detection than curve (b).

application of a detector to images containing lesions of the same type, but with different characteristics. Clearly, the detector is much better at detecting lesions in group (a) than in group (b). *Figure 11* also illustrates that the shape of an empirical FROC (or ROC) curve can deviate substantially from the idealized form shown in *Figure 10*.

8. Available material

8.1 Software

A great deal of software exists for different applications that have been discussed in this chapter. The following is a selection from my personal experience. It is by no means an exhaustive list, and a different author may have come up with a different set of examples. By the nature of things, these suggestions are likely to go out of date. A search of the Internet would almost certainly produce some useful (and even free) packages. I provide a few starting points for searching.

There are commercially available packages that allow data analysis to be performed without the requirement to do any programming. SPSS and SAS are both large data analysis packages which support (among many other functions) the calculations for statistical pattern recognition. They are available from SPSS Inc., 2333 S. Wacker Drive, Chicago, IL 60606–6307

4: Pattern recognition: classification of chromosomes

(www.spss.com) and the SAS Institute Inc., SAS Campus Drive, Cary, NC 27513–2414 (www.sas.com) respectively. Both of these organizations have distributors in several other countries. NeuralWare Inc., 202 Park West Drive, Pittsburgh, PA 15275 (www.neuralware.com) sell packages for building neural network applications. Their product Aspen NeuralSIM replaces an earlier package called Predict, which allowed, among other things, genetic algorithm search for the most discriminating feature combinations.

Potentially, more flexibility can be obtained by programming one's own solutions. In such cases, it may be useful to make use of high-level mathematical programming environments such as MATLAB from The Mathworks Inc., 24 Prime Park Way, Natick, MA 01760–1500 (www.mathworks.com). MATLAB provides not only a programming language with a syntax designed for mathematical calculation, but also a large number of 'toolboxes', providing comprehensive libraries of functions in different application areas, such as statistics, neural networks, and optimization. A great deal of third-party software is available for MATLAB, either free or at a price. Approved packages are listed on the Mathworks web site. Of particular interest in this context is a (currently) free package from the Neural Computing Research Group at Aston University in the United Kingdom (www.ncrg.aston.ac.uk/netlab/over.html). Netlab is a MATLAB package with functions for building a number of different neural network architectures as well as other pattern recognition procedures such as k-means clustering, k-nearest neighbour classification, and the EM algorithm. MATLAB implementations of genetic algorithms can be obtained from www.ie.ncsu.edu/mirage/ and from www.shef.ac.uk/uni/projects/gaipp/gatbx.html.

A free suite of programs for neural network simulation—the Stuttgart Neural Network Simulator, SNNS, can be obtained from the University of Stuttgart by anonymous ftp (ftp.informatik.uni-stuttgart.de) in directory pub/SNNS. These programs are only available for Unix systems.

It is often extremely useful to browse multidimensional data to get an informal impression of how features are correlated or clustered, which features are useful, and which are not. A useful tool for doing this on Unix systems is Xgobi. The origin of the name is obscure. The 'X' refers to the X windows system into which the software is integrated. It is available from www.research.att.com/~andreas/xgobi/ and further information may be obtained from Bellcore Customer Service, 8 Corporate Place, Piscataway, NJ 08554, USA.

There are a number of web sites providing introductory or more advanced statistical material. A useful site to start from is the CTI statistics site at the University of Glasgow (www.stats.gla.uk).

8.2 Further reading

The field of pattern recognition is quite wide. The material presented in this chapter has been introductory and has concentrated on practical issues. More detailed treatments can be found in a number of texts. Gose *et al.* (2) provide a useful introduction, covering the basic topics thoroughly with many worked examples and exercises. They concentrate on pattern recognition, with some space given to image analysis. Duda and Hart (1) is the classic text and is very good on basic methods, Bayesian statistics in particular. Therrien (4) provides a discussion of some topics not covered by the others at a straightforward mathematical level. Ripley (6) provides an up to date and comprehensive account, relating statistical methods to neural networks, while Bishop (7) concentrates particularly on the use of neural networks in classification. McMillan and Creelman (49) is a good source for detection theory and is well illustrated with examples. More advanced techniques, such as genetic search and the EM algorithm, are covered by Mitchell (9).

All of these books take their audience to be engineers or mathematicians at different levels, and assume some fluency in mathematical analysis. Ripley's book in particular is not for the mathematically timid.

Acknowledgements

I am grateful to Dr Susan Astley and her colleagues for permission to make use of the data from which *Figure 12* was constructed and to Dr Neil Thacker for comments on the manuscript.

References

1. Duda, R. O. and Hart, P. E. (1973). *Pattern classification and scene analysis.* Wiley, New York.
2. Gose, E., Johnsonbaugh, R., and Jost, S. (1996). *Pattern recognition and image analysis.* Prentice Hall PTR, Upper Saddle River, NJ.
3. Hand, D. J. (1981). *Discrimination and classification.* Wiley, Chichester, UK.
4. Therrien, C. W. (1989). *Decision estimation and classification.* Wiley, New York.
5. Tou, J. T. and Gonzalez, R. C. (1974). *Pattern recognition principles.* Addison-Wesley, Reading, MA.
6. Ripley, B. D. (1996). *Pattern recognition and neural networks.* Cambridge University Press.
7. Bishop, C. M. (1995). *Neural networks for pattern recognition.* Clarendon Press, Oxford.
8. Dempster, A. P., Laird, N. M., and Rubin, D. B. (1977). *Journal of the Royal Statistical Society, Series B*, **39**, 1.
9. Mitchell, T. M. (1997). In *Machine learning* (ed. C. L. Liu) (McGraw-Hill Series in Computer Science). McGraw-Hill, New York.
10. Dasarathy, B. V. (1991). *Nearest neighbor norms: NN pattern classification techniques.* IEEE Computer Society Press, Los Alamitos, CA.
11. McKay, R. J. and Campbell, N. A. (1982). *British Journal of Math. and Stat. Psychology*, **35**, 1.
12. Klecka, W. R. (1984). *Discriminant analysis.* Sage Publications.

4: Pattern recognition: classification of chromosomes

13. Davies, L. (1991). *The handbook of genetic algorithms*. Van Nostrand Reinhold, New York.
14. Goldberg, D. (1989). *Genetic algorithms in search, optimisation and machine learning*. Addison-Wesley, Wokingham.
15. Aarts, E. and Korst, J. (1989). *Simulated annealing and Boltzmann machines*. Wiley, New York.
16. Manly, B. F. J. (1986). *Multivariate statistical methods: a primer*. Chapman and Hall, London.
17. Sammon, J. W. (1969). *IEEE Transactions on Computers*, **18**, 401.
18. Lerner, B., Guterman, H., Aladjem, M., Dinstein, I., and Romem, Y. (1998). *Pattern Recognition*, **31**, 371.
19. Forgy, E. W. (1965). *Biometrics*, **21**, 768.
20. MacQueen, J. (1967). In *Fifth Berkley Symposium on Mathematical Statistics and Probability* (ed. M. N. Le Cam and J. Neyman). University of California Press.
21. Hall, D. J. and Khanna, D. (1977). In *Statistical methods for digital computers* (ed. K. Enslein, A. Ralston, and H. S. Wilf), Vol. 3, pp. 340–373. Wiley, New York.
22. Beale, R. and Jackson, T. (1990). *Neural computing: an introduction*. Adam Hilger, Bristol.
23. Dayhoff, J. (1990). *Neural network architectures, an introduction*. Van Nostrand Reinhold, New York.
24. Hagan, M. T., Demuth, H. B., and Beale, M. (1996). *Neural network design*. PWS, Boston, MA.
25. Fiesler, E. and Beale, R. (1997). *Handbook of neural computation*. Institute of Physics and Oxford University Press, New York.
26. Granlund, G. H. (1976). *IEEE Transactions on Biomedical Engineering*, **23**, 182.
27. Groen, F. C. A., ten Kate, T. K., Smeulders, A. W. M., and Young, I. T. (1989). *Pattern Recognition Letters*, **9**, 211.
28. Granum, E. (1982). In *Pattern recognition theory and applications* (ed. J. Kittler, K. S. Fu, and L. S. Pau), p. 373. Reidel, Dordrecht.
29. Piper, J. and Granum, E. (1989). *Cytometry*, **10**, 242.
30. Errington, P. A. and Graham, J. (1993). *Cytometry*, **14**, 627.
31. Errington, P. A. and Graham, J. (1993). In *IEEE International Conference on Neural Networks* (San Francisco, CA), p. 1236. IEEE, Los Alamitos, CA.
32. Lerner, B., Guterman, H., Dinstein, I., and Romem, Y. (1995). *International Journal of Neural Systems*, **6**, 359.
33. Lerner, B. (1998). *IEEE Transactions on Systems Man and Cybernetics, Part B—Cybernetics*, **28**, 544.
34. Nivall, S. (1995). PhD thesis, Chalmers University of Technology.
35. Lerner, B., Guterman, H., Dinstein, I., and Romem, Y. (1995). *Pattern Recognition*, **28**, 1673.
36. Piper, J. (1987). *Signal Processing*, **12**, 49.
37. Kleinschmidt, P., Mitterreiter, I., and Piper, J. (1994). *ZOR—Mathematical Methods of Operations Research*, **40**, 305.
38. Charters, G. C. (1994). PhD thesis, University of Manchester.
39. Charters, G. C. and Graham, J. (1999). *Pattern Recognition*, **32**, 1335.
40. Piper, J. (1994). *12th International Conference on Pattern Recognition* (Jerusalem, Israel), p. B529. IEEE Computer Society Press, Los Alamitos, CA.
41. Tso, M. K. S. and Graham, J. (1983). *Pattern Recognition Letters*, **1**, 489.

42. Tso, M. K. S., Kleinschmidt, P., Mitterreiter, I., and Graham, J. (1991). *Pattern Recognition Letters*, **12**, 117.
43. Efron, B. (1982). *The jackknife, the bootstrap and other resampling plans*. Monograph CBMS-NSF 38, Society for Industrial and Applied Mathematics, Philadelphia.
44. Efron, B. and Gong, G. (1983). *The American Statistician*, **37**, 36.
45. Efron, B. and Tibshirani, R. J. (1993). *An introduction to the bootstrap*. Chapman and Hall, London.
46. Cho, K., Meer, P., and Cabrera, J. (1997). *IEEE Transactions on Pattern Analysis and Machine Intelligence*, **19**, 1185.
47. Piper, J. (1991). *Pattern Recognition Letters*, **13**, 685.
48. Graham, J. and Errington, P. A. (1999). In Artificial Neural Networks in Biomedicine (ed. P. Lisboa, E. C. Ifeachor and P. S. Szczepaniak), Springer, Heidelberg.
49. McMillan, N. A. and Creelman, C. D. (1991). *Detection theory: a user's guide*. Cambridge University Press.
50. McNeil, B. J., Keeler, E., and Adelstein, S. J. (1975). *New England Journal of Medicine*, **293**, 2121.
51. Zwiggelaar, R., Parr, T. C., Schumm, J. E., Hutt, I. W., Taylor, C. J., Atley, S. M. and Boggis, C. R. M. (1999). *Medical Image Analysis*, **3**, 1.

5

Three-dimensional (3D) reconstruction from serial sections

FONS J. VERBEEK

1. Introduction

This chapter describes the theory and application of three-dimensional (3D) reconstruction from serial sections. The technique has a long history, starting by sculpturing from sections via photomicrographs to a wax model and evolving to a virtual 3D model exclusively existing in computer memory. The latter method is, at least to some extent, automated and therefore referred to as computer-assisted reconstruction (CAR). CAR is commonly applied in the laboratory. However, computerizing the reconstruction process has implications for laboratory practice, and clear and unambiguous terminology is important and necessary. The biological problems to be solved with 3D reconstruction tend to be complex. Ideally, finding solutions is an interaction between the life sciences and computer sciences. The computer[1] is the means to solve the problems raised, serving as the objective assistant to the process, and solutions are formulated in terms of applied physics since the whole procedure depends on the formation and manipulation of the microscope image. Computer science provides the computational resources because 3D reconstruction requires effective algorithms for processing, as well as for data management. Interpretation of the data of course remains the exclusive domain of the biologist.

Two methodological aspects will be discussed; the problem of realignment of the sections and the problem of appropriate representation for computerized processing. A 3D reconstruction can be represented by different geometrical representations that are appropriate for the task in hand, for example measurement and visualization.

1.1 3D reconstruction in microscopy

The area of application is the field of biomedical research at the level of tissues and organs in which knowledge of the third dimension is essential for

---

[1] In this chapter *computer* refers to a desktop model of computer architecture and an operating system. The applications described have been implemented on an Intel-based PC and/or a Unix graphical workstation.

the interpretation and understanding of research results. For the examples presented, 3D reconstruction is applied in order to reveal the spatial distribution of gene expression in embryos during development. Gene-expression patterns may serve as molecular markers to elucidate the tissue remodelling underlying normal, as well as abnormal, development. The gene expression is studied using its gene products, i.e. mRNA and proteins, and also referred to as a *phenotype* of the object under study.

Three-dimensional reconstruction is concerned with the assembly of information obtained in a two-dimensional (2D) plane of known orientation into a coherent and comprehensive 3D set of data. This set of data is usually referred to as the *3D reconstruction* (see Section 2.1). The *a priori* knowledge of the orientation of the plane of sampling is essential for the ability to accomplish the 3D reconstruction (parallel sampling planes). As a method for 'artificial' synthesis of an object out of a set of object views, 3D reconstruction is widely applied in biomedical imaging and robot vision. To name a few important techniques that aim at reconstruction of an object into its original 3D state, one can think of reconstruction from bi- and/or multiplane projections, stereo imaging (1) or 3D reconstruction from consecutive parallel planes (2–11). Methods used to obtain a 3D reconstruction have in common that the method of input (imaging) dictates the plane of sampling. The applications described in this chapter are based on sampling of an object into coplanar 'physical' sections; by elaborating on the possibilities for sampling into coplanar sections the choice of the sectioning technique will be justified.

1.2 Why serial sectioning is necessary

Serial sectioning is used to probe the internal structure of an object that can not be investigated by other means. In this section the possibilities of studying the internal structure of an object are examined.

1.2.1 Physical versus virtual sectioning

With respect to scanners producing consecutive (coplanar) sections, the resolution in the direction of sectioning, referred to as the z-resolution, is related to the size of the object under study. In routine applications of computerized axial tomography and nuclear magnetic resonance (NMR) imaging, z-resolution is of the order of millimetres. A confocal laser scanning microscope (CLSM) can produce optical sections of the order of tenths of a micrometre (4). For 3D imaging applications where a z-resolution of (tens of) micrometres is required, coinciding with the resolution required for 3D studies in the field of microscopical anatomy and developmental biology, other techniques are necessary.

In contrast to virtual sectioning techniques (non-invasive), acquisition of z-samples in the range of (tens of) micrometres, physical sectioning (invasive) is used with the obvious disadvantage of loss of integrity in the third dimension

5: Three dimensional (3D) reconstruction from serial sections

combined with a possible deformation of the object in the section. The 3D relations can only be restored to their native state by application of 3D reconstruction.

In NMR microscopy new developments have led to increased z-resolution as a result of which it is possible to scan specimens at the level of a small vertebrate foetus (12) or embryo (13). However, in contrast to invasive sectioning, application of NMR microscopy does not allow probing of internal structures of the object by means of their phenotype. This is because NMR imaging is based on a physical principle, the magnetic moment of, for example, hydrogen in water, whereas imaging of a molecular phenotype, that is a genomic marker, requires using histochemistry to obtain in situ staining. Hence, for studies with an emphasis on patterns of gene expression this is not the most appropriate technique.

CLSM allows imaging of internal structures using fluorescent probes. Accurate optical sections are produced, but optical penetration into the specimen is limited. The scope of CLSM is on the cellular level rather than on that of an organ or a tissue. Consequently, for 3D reconstruction of an embryo or a substantial part of it, CLSM is usually not the most appropriate technique.

Invasive sectioning enables the application of staining techniques so that molecular phenotypes of the specimen under study can be revealed. The production of 3D reconstructions from serial sections requires a technique to correct for loss of alignment. If shape description is pursued, a possibility to correct for the deformation introduced by the sectioning should be included in the reconstruction process. Several approaches for the 3D reconstruction from serial sections have been described ranging from hand-made 3D reconstructions (2, 3) to semi- and/or fully automated methods (5, 7–9, 14, 15–17).

1.2.2 Measurements

Assuming accurate 3D reconstructions can be obtained, they will enable 3D measurements as well as visualization of the restored third dimension. Ideally, correct 3D reconstruction is independent of the direction of sampling. The question arises as to whether the availability of a 3D reconstruction is essential for 3D measurements. In some cases 3D measurements can be accomplished using stereology (18). However, correct application of stereology usually requires that assumptions on the shape can be made beforehand. Moreover, stereology is a method to obtain measurements from samples of objects by statistical means resulting in average values of the samples, but not their spatial distribution. A numerical description of the shape of patterns of gene expression is excluded from this approach, since having no assumptions on the shape of the patterns before having brought them into their native state is the starting point. It should also be stressed that spatial shape parameters can only be extracted from individual 3D reconstructions. The more complex are the structures under reconstruction, the more important it is to visualize 3D relations. This feature, in particular, contributes to the popularity of application of 3D reconstruction in developmental biology studies.

1.2.3 Application of 3D reconstruction to new data

If a new reconstruction from a serial section is planned, a number of items should be discussed and checked:

Protocol 1. Preliminary considerations at the onset of making a 3D reconstruction

Nature of the data (input)
- Choose the embedding medium to suit the required resolution and sectioning method.
- Determine the required extent, scale, and resolution of the data.
- Establish the required contrast of the data (i.e. what tissues/signal must be visible), and whether this can be applied with the embedding medium.
- Determine what accuracy is required of the method considered.

Nature of the result (output)
- What information should the reconstruction convey?
- Is graphical presentation of the data sufficient, i.e. simply visualization?
- What measurements are to be made on the reconstructed data?
- What are the requirements for computing facilities (storage and CPU power), i.e. are the current facilities sufficient?

In the next sections the methodology to be able to assess this checklist and apply reconstruction will be discussed.

2. Methodological aspects

It is obvious that due to size, probing internal structure is done by means of microscopy. Serial sectioning is accomplished by microtomy and results in the cutting of a biological specimen into parts. In biomedical applications these sections are prepared and studied with various histological techniques. Consequently, the morphological distortion, which is inherently related to physical sectioning of histologically prepared material, has to be understood and addressed. The problems arising as a consequence of the sectioning—the distortions—are described.

A distortion is an alteration of the original geometrical configuration of the object and the object sample, the section. Distortions are described in terms

5: Three dimensional (3D) reconstruction from serial sections

of geometrical transformations imposed on the object. The following distortions are inflicted on the object and/or section:

- a destruction of 3D integrity
- a distortion due to compression at cutting
- a distortion resulting from the mounting process
- a mirror reflection by improper mounting of the section

Physical sectioning is a sampling of the object in the z-direction. A direct result of this process is the loss of the integrity of the object in the third dimension, specifically in the direction of sampling (*Figure 1A*). Interpreting distortion in geometrical terms means, relative to the object in the embedding block, the section changes in position and orientation (*Figure 1A,B*) or, in mathematical terms, a translation and a rotation, respectively. Next, while the section is cut, a distortion resulting from a compression perpendicular to the direction of sampling develops. This deformation is a distortion of proportions which relates to scaling in mathematical terms, i.e. a combination of tensile and compressive stress.

Sections are mounted on slides for further histo-chemical treatment. Compression introduced by the sectioning process ought be removed during the mounting step (*Figure 1B*). This is achieved by applying the section to a water film on the slide. Apolarity of the embedding material causes the sections to decompress over the water film. At the same time other distortions may be introduced if the stretching causes decompression in different directions. In the worst case these decompressions are unevenly distributed (*Figure 1B*).

The type of embedding material, together with the section thickness,

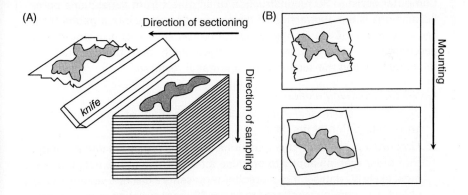

Figure 1. (A) Schematic drawing of the process of sectioning with a microtome. The knife cuts through the embedding block containing the object and the resulting section is compressed in the direction of the sectioning. (B) Schematic drawing of the process of mounting. The section decompresses, eventually, decompression occurs unevenly in different directions.

determines the extent of the deformation inflicted. In general, the softer the embedding material, the more severe the extent of the deformation, and also, the thinner a section is, the more it is subject to deformation.

Occasionally, a section may flip during manipulation and consequently the section is mounted upside down on the slide. In an automated reconstruction system it should therefore be anticipated that mirror reflection might have occurred. Flipping of sections typically occurs with plastic sections that are picked up one at a time. In general, wax sections are picked up as ribbons and thus are not so susceptible to inversion.

Basically, reconstruction from serial sections requires: (1) a procedure for realignment (including deformation correction), and (2) a procedure for data management. It is good policy to start with good data management, meaning the images in the reconstruction process are retrieved from a database that also keeps track of all data additional but indispensable to the process. In this manner restoration to a previous state can be conveniently pursued. For 3D processing of the reconstructed data, i.e. measurement and visualization, such a database will be of great value.

2.1 Modes of action

In the production of a computerized 3D reconstruction different modes of interaction with the data are distinguished, and in doing so, insight is obtained on manipulation of the data at each point in the reconstruction process (8, 11):

Box 1. Computer-assisted 3D reconstruction

Definition

Computer-assisted 3D reconstruction of an object from serial plane-parallel sections is the abstraction of the series of sections into a *geometrical representation*. Subsequently, a pile of consecutive sections stored in a geometrical representation is rearranged such that the original 3D relations between the sections are (optimally) restored.

Levels of data manipulation

1. *Zero-order processes*—manipulations consisting of the acquisition and organization of the data.
2. *First-order processes*—manipulations using the organized dataset of the zero-order manipulations to produce a reassembly of the data into the most likely 3D dataset that thereafter represents a discrete version of the object. This version is referred to as the 3D reconstruction of the object.
3. *Second-order processes*—manipulations using the dataset that has been produced by the first-order manipulations to extract information from the dataset.

2.2 Organizing the reconstruction data

Organization of data required for the reconstruction should start in the acquisition phase. Essentially, 3D reconstruction is a simulation of a real-world object. This object is sampled into sections and each section is simulated by a geometrical representation which in the digital sense it is an *image* and/or a *contour*, both obtained by a digitization process in the *xy*-plane (*Figure 2A,B*). All sections are strictly plane-parallel and, consequently, the direction of sectioning coincides with the *z*-axis. In this manner a system of three orthogonal axes is obtained (see Section 2.4).

2.2.1 Relation between section and image

Because the image is related to a physical section of known thickness, the *section-image* is introduced. This term is explicitly used in the case that an image represents a section.

2.2.2 Modelling 3D sectioned biological data

A comprehensible model is proposed to symbolize section data in an efficient manner. This model of the data should be convenient for further processing, i.e. visualization and analysis referred to as second-order manipulations (see Section 2.1). The *object* is the key entity in the model; it represents the shape to be reconstructed (6–8, 11). The object consists of several *structures* that can be annotated in the section-image(s); each of the structures itself represents a shape. Next, an object is sampled into *sections*, represented by section-images, and each of the section-images contains structure intersections. In this manner the object can be accessed either by a structure or by a section.

The structure intersection entity is conveniently expressed by a contour or by the area delineated by a contour. A contour is a polynomial represented by a list of *xy*-coordinates. In *Figure 7*, examples are given of such annotations. Use of images in the reconstruction process (see Section 2.2.1) will accomplish a coherent true 3D stack of images that can be visualized as such (see *Figures 7* and *9*). This representation of the density voxel model is used with the labelled voxel model and the contour model (11).

2.2.3 Application of the 3D object model in the acquisition phase

The above model is implemented in refs 6–8 and 11; each of the entities holds information specific to that entity and duplication of information is prevented. Moreover, the model is implemented such that different geometrical representations can be used.

Structures are added to the database having a unique name, thereafter on the level of the user the structure is accessed by name. Likewise, a sequence of sections is entered into the database and thereafter, sections are accessed by sequence number. Structure intersections (contours) are addressed by combining structure name and section number. Addressing data in the database is

Fons J. Verbeek

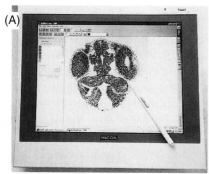
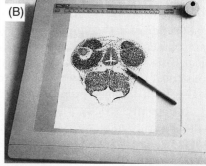
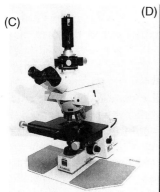
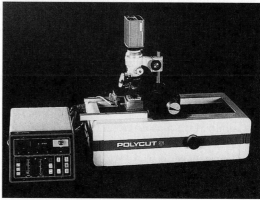
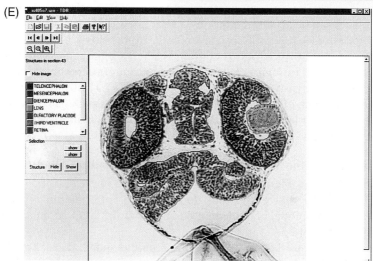

5: Three dimensional (3D) reconstruction from serial sections

Figure 2. (A) Digitizer tablet with cordless stylus. A photograph is mounted on the tablet surface for contour drawing. (B) LCD tablet with cordless stylus and section image to annotate rendered on the LCD screen. (C) Microscope equipped with a computer controlled *xy*-stage and rotatable camera mount. The camera mounted is a standard research CCD camcera. (B) Horizontal microtome equipped with a bridge on which a epi-illuminated microscope is mounted. Images are taken from the block as the object stops in the centre of the field of view of the CCD camera (see *Figure 1*). (E) Layout of the interface for the annotation of images by contours. The contours can be used in a contour model. The interface shows how a selection of structures (contours) can be made from the geometrical database for further processing.

---

accomplished by queries which are provided in a predefined form, e.g. as menus and selection boxes (*Figure 2E*).

Geometrical representations with the object model

The model is described using the contour as the basic geometrical entity (see Section 2.1); this is, however, not always the most practical representation. Therefore, the model is extended to be independent of geometrical representation. In the input phase, apart from the contour, a labelled area (voxel plane) can be used as geometrical structure. Comparable to the contour, each labelled area in the section-image represents a structure intersection (with the section).

Once the step to a 3D model is made, other representations come into view. Representations to describe the 3D reconstruction are: contour stack, voxel,[2] and surface patch model. Each of these representations has its own advantages (6, 11). If, initially, a reconstruction was prepared as a contour model, the others can be derived by conversion.

2.3 Methods and devices for the input of the data

In modern usage of 3D reconstruction from serial sections, especially in automated applications, input is typically restricted to two devices, namely a camera[3] and a digitizer tablet. The camera is mounted on the microscope to acquire images of each section or part of it. The digitizer is used to get polynomial information out of drawings or photographs mounted on the digitizer surface. Additionally, the digitizers are used as accurate drawing devices to manually indicate relevant boundaries in images (*Figure 2A,B*).

2.3.1 Image acquisition

In addition to the microscope (compound research microscope) and computer, a basic image acquisition system requires three more components: the camera, the framegrabber, and the software to support the control of camera and framegrabber. Inoue and Spring (19) have published a comprehensive

---

[2] The voxel model is also known as a volume enumeration model.
[3] CCD technology is commonly applied in the laboratory; in this text a reference to a camera implies the use of a CCD camera, unless specified otherwise.

report of the use of camera systems in microscopy, including detailed descriptions of the mode of operation of the cameras (see Chapter 1). For digitization of the video signal a framegrabber is required.

With respect to the acquisition all precautions to obtain images with even illumination should be taken into account at the onset of an acquisition session. This is important as shading might have an (unwanted) effect on the outcome of the transformation estimation, e.g. in a segmentation procedure or in a matching of grey-values (see Section 3).

As images are used to estimate geometrical transformations it is important to be sure of correct sampling by the imaging system of geometries present in the section, i.e. no introduction of (spherical) aberrations. This needs measurements at the set-up of the imaging system by using a calibration slide which consists of a set of known shapes with known dimensions. These measurements should be completed for all lenses in the imaging system (see *Protocol 2*).

Camera and framegrabber

Using the resulting 3D model for measurements requires sampling of the *xy*-plane with the highest possible accuracy, taking the lens properties into account. A camera with square sensor elements (isotropic sampling) is preferred. As image densities must be consistent throughout acquisition of a series of images the combination of camera and framegrabber should be configurable such that the user has control of the video signal (see Chapter 1).

Image acquisition software

Image acquisition on a series of sections requires a mode of sequential operation. In general, standard packages for image acquisition do not include efficient control for the acquisition of a sequence of images. Moreover, having available additional parameters of the 3D reconstruction is convenient in further processing (e.g. section thickness, lens, auxiliary lens). To this end, in advanced systems 'home-made' software is used typically. Advanced acquisition systems for 3D reconstruction use a database to keep track of the abundance of data in support of a 3D reconstruction. In Section 3 two image acquisition systems embedded in the methodical approach taken for the 3D reconstruction and which incorporate a database are discussed.

Protocol 2. Image acquisition for serial sections

Materials
- Image analysis micrometer (Edmund Scientific, #F53713, ± $300) www.edsci.com
- Test slides

Checklist to consider
- Determine the number of sections on a slide (max. 20); this is preferred for efficient processing (see Section 4.1.).

5: Three dimensional (3D) reconstruction from serial sections

- Code the slides from A01 to A##. Sections are numbered as A01.01–01.20, etc.
- Determine a stepsize, i.e. every section or every other section (or large steps if necessary).
- Establish the resolution required, i.e. the finest detail visible in the reconstruction.

Initial steps

1. Determine the objective lens required to achieve the required resolution.
2. Use the test slide to calibrate the image geometry.
3. With required lens do test acquisitions so as to obtain good contrast (colour filters); the image histogram can be addressed for objective evaluation of contrast.
4. Choose an image size that comprises the field of view (FOV) and resolution.
5. Check that the largest object part still fits in the FOV.

Method

1. Start acquisition with largest object in the series.
2. The first image is numbered an arbitrary high number, e.g. 500.
3. The next image is numbered 500 + section step; e.g. 502: ,<name>500, <name>502, etc.
4. Keep track of adjustments to the imaging system per section-entry.
5. Write a comment per section using unique ID.
6. After acquisition the sequence can be renumbered taking into account the total number of sections in the series in such a way that the image sequence can be extended.

Suggestions for file format

It is recommended to store data in a text file (20) that can be read by all kinds of programs. Tab-separated files are easy to used and can be read in Excel. For example:

<number><tab><47><CR>   number of sections used in this reconstruction
<thickness><tab><7><CR>   section thickness, here indicated in micrometres
<lens><tab><20><CR>   lens used, may point to a table with pixel sizes
<filter><tab><ND50><CR>   neutral density filter used

Protocol 2. *Continued*

For the acquired sections a comparable format can be utilized. For example:

\<A01.15\>\<tab\>\<lamp adjusted to 7.5\>\<CR\>
\<A02.08\>\<tab\>\<not used, bad section\>\<CR\>
\<A07.10\>\<tab\>\<staining weak, used green filter to obtain more contrast\>\<CR\>

If necessary all images can be easily traced back and re-acquisition can be considered. Information can be stored in different files. The relation between the files can be documented in one of the files.

2.3.2 Acquisition of contours

The acquisition of contours is accomplished with a digitizer tablet connected to the RS-232 port of the computer. The digitizer consists of a tablet and a pointing device, i.e. stylus or puck. Under the tablet surface sensors are mounted that allow rapid conversion of the *xy*-coordinates indicated with the pointing device on the digitizer surface. The generic digitizer has an active surface of 30 × 30 cm and a resolution of 0.1 mm can be achieved (see *Figure 2A,B*).

Digitizer tablets

Digitizer tablets are used in image analysis because manual tracings and points of interest can be more easily accomplished. Image analysis software in the late 1970s and early 1980s integrated these devices in their user interface; the whole of the program functionality was mapped to the active area of the digitizer tablet. Software drivers were included in the application software. Standardization of operating systems has resulted in full support of digitizer tablets (e.g. MS Windows 3.x; 9x and NT, Solaris, Irix and MacOs) by means of system software drivers. Digitizers are modernized in the sense that the digitizer stylus is made cordless which results in a more convenient drawing operation. The digitizer stylus combination for manual tracing remains superior over mouse devices for its precise control over the drawing process.

LCD digitizers

The LCD tablet is a digitizer device integrating annotation of images. The digitizer surface of this tablet is at the same time a computer display so that instead of working with a pointing device and requiring trained eye–hand coordination for a fast drawing/annotation process, the feedback is direct. This tablet is equipped with a cordless pen and has all the advantages of the digitizer, being the preferred drawing device.

5: Three dimensional (3D) reconstruction from serial sections

---

Box 2.   Tips on digitizers

Digitizers frequently used, and for which drivers are available:

- Genius   NewSketch 1212   www.genius-kye.com
- Summagraphics   Bitpad II   www.summagraphics.com
- Wacom   Digitizer II, LCD PL400 www.wacom.com

The URLs are given to secure up to date information.

Contour acquisition software

An object consisting of several structures results in several groups of contours. This information needs to be efficiently managed and maintained. The proposed model for sectioned biological data allows such maintenance (7). In the applications described in this chapter, the acquisition of the contours is organized around the geometrical database. At input a section (section-image) and a structure are selected and contours are drawn using the digitizer tablet. At this point entries in the database are filled and a list of contour points is stored (polyline).

Box 3.   Software packages

Contour annotation

Packages for making annotation in images by contour drawing are most often commercial. From a prototype (7) an improved version of our 3D reconstruction software is currently being implemented. A trial version for annotation and management of contour data will be available for evaluation and can be downloaded from www.niob.knaw.nl (imaging).

Image processing packages

For acquisition of sequences of images packages with framegrabber support and the availability of writing scripts:

Scion Image (NIH-Image) Scion Inc., USA   www.scioncorp.com
SCIL Image   TNO, The Netherlands   www.tno-tpd.nl

For processing sequences of images the following is also of interest:

CVIP Tools   University of Southern   www.ee.siue.edu/
                         Illinois   CVIPtools

These are packages that the author has used successfully. They are well documented and (at least) trial versions are available from the URLs. Scion-Image is a dedicated version of NIH-Image.

2.4 Spatial resolution

For quality in producing measurements and visualizations from the 3D reconstruction the best possible spatial resolution is required.

2.4.1 Spatial resolution in the *xy*-plane

Using section-images as a starting point for 3D reconstruction, the resolution of the microscope lens, namely numerical aperture (NA), in combination with the camera determines the sampling in the *xy*-plane. Ideally, the camera has square sensor elements with little or no spacing between the sensor elements to guarantee optimal sampling. If the sensor does not have square sensor elements resampling should be considered for convenient measurement and visualization.

2.4.2 Spatial resolution in the *z*-direction

Section thickness as adjusted on the microtome determines the sampling along the *z*-axis. In this manner physical sectioning imposes the most severe constraints on the resolution of the final dataset. The sampling in the direction of sectioning, however, is always bound by physical limitations of the sectioning, the price for probing internal structures. Therefore, it should be taken into account not to expect more in computer memory than is sampled at a specific resolution. Higher sampling density in the *xy*-plane, using a lens of higher magnification (higher NA), results in more anisotropic sampling with respect to the *z*-direction. This will show as discontinuities in the 3D reconstruction and might be solved by interpolation which, however, does not add any new information to the reconstruction.

2.4.3 Datastructure and resampling

It is preferred that a 3D image has isotropic image elements. In 3D, cubic voxels will allow the straightforward application of digital measurement and filtering. Using standard lenses, combined with a workable section thickness, it is obvious that if 3D images consist of isotropic voxels, resampling according to a given section thickness is required. All of the information necessary to produce isotropic voxels is available from the datastructure used in the input phase. Here the particular advantage of a well-defined datastructure becomes apparent. If resampling is applied, all high-resolution data can be kept and used upon request. Annotation of the image is preferably done in the images with the best resolution; after annotation the contours and the images are rescaled so that, for routine applications, an isotropic image is used.

2.5 Alignment and deformation correction

Section 2.3.1 explained how a sequence of section-images is obtained. This transfers to a real 3D image assembling all section-images in the stack. Unfortunately, the 3D integrity and thereby the original 3D relations are

5: Three dimensional (3D) reconstruction from serial sections

destroyed by the sectioning process. Therefore, a procedure has to be designed to restore the 3D relations as closely as possible.

2.5.1 Semantics and definitions

The effect of the sectioning process is well described by a possible change in position and orientation of the section with respect to the z-axis of the object. The procedures used to restore the set of section-images to a valid 3D image are referred to as first-order manipulations (see Section 2.1). In order to be able to reassemble a stack of section-images with respect to a hypothetical z-axis, a relation between the section-images needs to be established. This relation is established by the process of *image matching*. For computerized 3D reconstruction purposes:

> Image matching is the process of determining and/or measuring correspondence between images by comparing mutually detectable features, e.g. pixels, lines, arcs, regions (21).

In 3D reconstruction the result of a matching operation between two images (or features thereof) is used to reflect a measure of the difference between the images. Since the differences in images are caused by processes that can be modelled by geometrical transformations, this process is used to bring the disparate images into correspondence. In the reassembly process of 3D reconstruction, the matching is usually pairwise and the matching features are chosen so that parameters for geometrical transformation can be derived. The process that aligns one image with respect to another image is referred to as *image registration* (21, 22):

> Image registration is the process of overlaying two images of the same scene by translation(s) and rotation(s), so that all corresponding points of the two images are coincident with one another.

The process of registration has to be applied to the whole range of images present in the geometrical representation and is referred to as *realignment*:

> Image realignment is the application of matching and subsequent registration over a stack of section-images of an object in order to restore the integrity of the 3D relations in the direction of sampling.

The reassembly is accomplished by a realignment procedure over the section-images. From previous definitions it is seen that the 3D integrity distortion is removed by rotation and translation while deformations (proportional-distortion) were also distinguished. Therefore, if applicable, proportional corrections should to be part of the reassembly procedure. The process that describes all of the transformations required is known as *congruencing* (21):

> Congruencing is the process of geometric transformations by which two images (or their features) of different geometries, but representing the same scene and/or object, are spatially transformed so that size, shape, position, and orientation of the scene/object in one image is made to be the same with respect to the size, shape, position, and orientation of the scene/object in the other image.

The process of congruencing accomplishes positional as well as orientational correction; therefore congruencing is, at least, an extension of image registration. In cooperation with matching, congruencing may be understood as a process of geometrical fitting of shapes. The extent to which a shape fits is determined by matching of these shapes. It should be noted that, with respect to 3D reconstruction, congruencing can only be applied in a limited amount of methodical approaches, i.e. an absolute structure of reference should exist (see Section 4.2).

The distortions under consideration are a result of the preparation method and should not be confused with distortions from the imaging device. Consequently, the image is only restored to the most probable situation before the object is subjected to the sectioning process. The section-image resulting from the first-order manipulation simulates as if distortion by sectioning had never happened.

2.5.2 Fiducials and estimation of transformation

The process of reassembly is controlled by criteria relating one section-image to a pivot section-image in the stack. Commonly, but not necessarily, the pivot section-image is an adjacent section-image. Several structures of reference, based on different features in the section-image, are used to explicitly establish the geometrical relation between section-images by application of a matching procedure that quantifies the difference between section-images so as to find a geometrical transformation minimizing the difference between these structures. Success of registration and/or deformation correction is related to the accuracy of matching that in turn depends on the fidelity of structures of reference. It is assumed that structures of reference are available for consecutive section-images over the image stack.

Structures of reference are referred to as fiducials. On the basis of their use in the matching procedure, fiducials can be divided into two groups: point-oriented and region-oriented. The point-oriented fiducials (point fiducials) are most often deduced from markers that are in some way distinguishable in the image. Using the image or a substantial part of it in the matching procedure implies the use of region-oriented fiducials (region fiducials).

5: Three dimensional (3D) reconstruction from serial sections

Point fiducials

Fiducial markers that are artificially added to the object are referred to as *reference points*. Otherwise, if a fiducial point is part of a structure belonging to the object and known *not* to change over a number of sections, it is referred to as a *landmark point*; e.g. an anatomical part.

Box 4. Point fiducials—assumptions and constraints

Assumptions
- All fiducials are in the field of view at acquisition.
- Corresponding point fiducials in the adjacent images are identifiable.
- Equal numbers are extracted from both images.

Constraints
- The artificial fiducial or reference point should be strictly perpendicular to the direction of sectioning; otherwise the oblique orientation should be known so as to avoid unjust straightening.
- A landmark point can be used if and only if the shape-change of the landmark is small.

In 3D reconstruction the concept of corresponding points is used as soon as the point fiducials are grouped in pairs. Pairs of corresponding points are referred to as a point-pattern. Point-pattern matching is a means of reducing the computational load and complexity of the matching problem and the subsequent computation of a geometrical fit (see Section 3.4).

Reference points are absolute fiducials if given constraints are properly taken into account, whereas landmark points can only be relative fiducials. Consequently, congruencing can be applied using reference points, whereas with landmark points one should limit to registration only. It is, however, a fact that sections may have been subject to different deformations thus introducing uncertainties in the computation of a registration/deformation correction as obtained from reference points. Likewise, the rate of change of landmark points may introduce errors in the transformation estimation thereby causing inexact registration.

Region fiducials: landmark shapes

A landmark shape is an enclosed region present in the section-image and representative for the object and/or the section under study. It is equal to the structure of reference. Typically landmark shapes are extracted from the section-image before use in a matching procedure which is usually realized through extraction of shape features.

> **Box 5.** Landmark shape constraints
>
> - The number of landmark shapes submitted to the matching is the same (preferred).
> - If only one landmark shape is available, it should not be bilateral symmetric.
> - If more than one landmark shape is used, at least one should not be bilateral symmetric.
>
> Using landmark shapes in adjacent sections and taking into account the condition of minimal shape change imposes extra constraints:
>
> - The relative position and orientation with respect to direction of sampling should be equal.
> - The relative proportions with respect to direction of sampling should be equal.
>
> The latter constraint especially applies if deformation correction is considered.

In contrast to point fiducials, region fiducials allow a better assessment of registration and/or deformation correction, i.e. measurement of the difference between fiducial region after transformation with respect to its reference.

In the case of a bilateral or *n*-fold rotationally symmetric shape the alignment is ambiguous and the correctness of the registration derived from the matching is doubtful. The addition of one landmark shape to break the symmetry can solve this problem.

The use of landmark shapes from adjacent sections for computation of a registration and/or congruencing requires caution. If ignored, each of the constraints imposed on the landmark shape may introduce an error that is amplified over the entire reconstruction. These errors are related to the nature of the 3D configuration of the shape in the unsectioned object from which the landmark shapes in the section-images are derived. Four (basic) classes are distinguished—the landmark shape is derived from a shape with:

I   Parallel orientation relative to the direction of sampling
II Oblique orientation relative to the direction of sampling
III A twisted nature
IV A tapered nature

The shape can also be a combination of either the classes I, II, III, and IV. Shape class I is the only legitimate shape from which landmark shapes can be derived. In *Figure 3* a number of simplified situations are depicted to illustrate incorrect 3D reconstructions.

5: Three dimensional (3D) reconstruction from serial sections

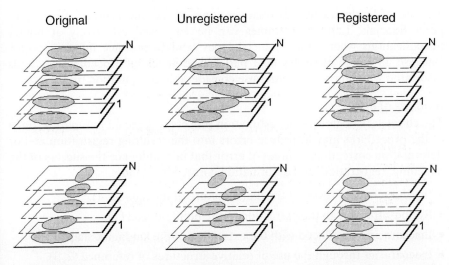

Figure 3. Effect of misregistration and false deformation correction as a result of wrongly chosen landmark shapes. In the upper row a configuration of an obliquely oriented landmark shape (before sectioning) is depicted, the stack of unregistered sections (after sectioning) that is to be registered according to the landmark shape in section 1. The third picture of this row depicts the result of incorrect application of a registration based on a set of landmarks not obeying the given constraints. The obliquely oriented cylinder is straightened and consequently the whole object under reconstruction is registered incorrectly. In the lower row a configuration of a twisted and tapered landmark shape is depicted, the unregistered set of sections resulting from sectioning this landmark. The third picture in this row represents misregistration of the twisted and tapered cylinder landmark structure. If, next to a registration based on the landmark shape of section I, a deformation correction is also applied, then this would result in loss of torsion as well as loss the tapering of the landmark structure.

Region fiducials: landmark frames (density data)

A *landmark frame* is represented by the section-image including the object under study, taking into account the intensity (grey-value) at every sampling point. Methods for matching landmark frames are related to cross-correlation of images (8, 9, 23, 24) and based on the intensity values rather than shape information.

Box 6. Landmark frame constraints

- Preparation (e.g. staining) of the sections should be the same.
- Conditions of image acquisition of the section-images should be more or less the same.

The landmark frame can be successfully used if:

- The extent of the deformation is not too large and mostly linear in nature.
- The distance between sections is not too large.
- Intensity fluctuations in the images are slowly varying.

The use of landmark frames removes some of the uncertainties ascribed to point fiducials. Landmark frames can be employed effectively at higher magnifications where other fiducial types tend to produce inferior results. Because all image elements take part in the matching procedure, methods associated with landmark frames use considerable computation time.

Sources of error

The characteristic flow of the estimation process is depicted in *Figure 4*. Each of the procedures may introduce errors into the resulting registration and/or deformation correction. Sources of error that may influence the success of the matching, registration, and deformation correction include:

- Inaccuracies as a result of the digital nature of the data (1–4).
- Round-off errors in the extraction of the point fiducial (1).
- Inaccuracies introduced with the extraction of the landmark shape (1).
- Inaccuracies through the use of relative structures of reference (2, 3).
- Inaccuracies as a result of the application of the wrong transformation (3, 4).
- Inaccuracies as a result of interpolation and a round-off error on a rectangular grid (4).

The digital nature of the images results in inaccuracies found in the whole process of the first-order manipulations. If the sampling is performed at correct magnification the effect can be kept minimal (25). Taking into account these inaccuracies, one should realize that an exact reconstruction only exists in theory. Depending on the fiducials and the nature of the fiducials an exact fit may be best formulated in terms of a confidence interval.

From evaluation of the fiducials it is obvious that without an absolute structure of reference, the application of deformation correction along with registration is illegitimate. Nevertheless, some techniques do incorporate deformation correction using relative structures of reference (26–28). A

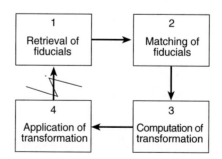

Figure 4. Flow diagram depicting the course of actions in first-order manipulations of 3D reconstructions from serial sections. After application of a transformation, the result may be used in an iterative fashion to achieve a best fit.

5: Three dimensional (3D) reconstruction from serial sections

registration derived from the matching of relative structures of reference suffers from the drawback of the introduction of an unknown error. Most methods (5, 24, 29) aim at the best possible registration. Similarly, the objective of the matching and the subsequent registration (and/or congruencing) on the basis of fiducials is formulated less stringently by an indication of the inexactness of the method, i.e. the best fit.

For accurate shape analysis, deformation correction is required. Concluding from the assessment of the fiducials and the sources of error, the premises of matching of adjacent sections may not be the best way for a reliable reconstruction of shape, as the main source of error will always be caused by the differences between two adjacent sections.

3. Mathematical aspects

A method for realignment requires assessment of the transformation applied. Realignment itself can be based on hardware whereas assessment of the fiducials is best dealt with in software. In this section the estimation of the geometrical transformation is discussed. Next, a number of assessment tools are formulated as well as how to invoke these tools in the realignment process. The assessment differs for each type of fiducial.

3.1 Estimation of transformations

The geometrical fit aimed at in the first-order manipulations is composed of a number of transformations that are favourably dealt with in one transformation. This is not always possible, and therefore the geometrical fit is disassociated into global and local congruencing transformations. The global transformations are described by linear transformations:

> A linear transformation is a mapping from one vector space V into another vector space V'. The following properties should apply for vectors **a** and **b**: $L(\vec{a} + \vec{b}) = L(\vec{a}) + L(\vec{b})$, where L indicates the linear operation; $L(n \times \vec{a}) = n \times L(\vec{a})$, where n indicates a scalar.

Application of registration (see Section 2.5.1) is equivalent to a rigid transformation. It supposes a body under rigid movement that translates and rotates, i.e. defined as:

> A rigid transformation $G(\vec{a})$ is a transformation that consists of the combination of a rotation $L(\vec{a})$ and a translation $\vec{t}$, with $\vec{a}$ a vector in Euclidean space.

If deformation correction is considered, a scaling and possibly skewing will need to be part of the transformation. The transform is then affine, which is a general linear transform with the properties:

1. Parallel lines are transformed into parallel lines.
2. Proportional distances are preserved, thus midpoints of lines remain midpoints.
3. Area scales as det|A|,[4] where **A** is the linear transformation matrix.

These statements are formally written as:

$$G(\vec{a}) = L(\vec{a}) + \vec{t}. \qquad [1]$$

Global congruencing transformations are described with an affine transformation applied in the *xy*-plane. The formal definitions are now written in coordinate space where a coordinate space (x,y) is mapped onto a new coordinate space (u,v), written as: $u = f(x,y)$ and $v = g(x,y)$. These functions are written as affine mappings:

$$f(x,y) = a_{11}x + a_{12}y + a_{31} \quad \text{and} \quad g(x,y) = a_{21}x + a_{22}y + a_{32}. \qquad [2]$$

In *Equation 1* it is seen that in linear transformations the translation is treated separately. In terms of matrix computations, and thus implementation, this is inconvenient. In order to be able to include the translation(s) directly into the matrix operations, homogeneous coordinates are introduced. In this section, the concept of homogeneous coordinates is used in all matrix operations. Written in matrix notation using homogeneous coordinates:

$$(u,v,1) = (x,y,1) \cdot \begin{bmatrix} a_{11} & a_{12} & 0 \\ a_{21} & a_{22} & 0 \\ a_{32} & a_{33} & 1 \end{bmatrix}, \qquad [3]$$

often simplified to

$$(u,v,1) = (x,y,1) \cdot \mathbf{A}. \qquad [4]$$

The process of estimation of a transformation can be schematized with *Figure 4*. First the landmarks are used to estimate a transformation, and then the transformation is accomplished. Next, the result is compared with the reference landmarks; the evaluation phase. The assessment of goodness of fit is used to decide whether a fine-tuning transformation with new parameters is required. This might imply iteration of the estimation process (see Sections 3.4 and 3.9). The matching depends on the type of landmarks combined with a priori knowledge of deformation of the embedding material.

3.2 Spatial transformation of image values

As the image is defined on a grid, a transformation may result in non-grid points. To this end a spatial transformation of the image is necessary. The

---
[4] det|L| refers to the determinant of the matrix **L**.

5: Three dimensional (3D) reconstruction from serial sections

spatial transformation is a mapping function that establishes a spatial correspondence between all points in an image and its transformed counterpart. The geometrical transformation is applied to an image which is then evaluated at each grid point. The result is again set to grid points by interpolation. A number of interpolation methods are available, including nearest neighbour interpolation, bilinear interpolation (30, 31), and spline interpolation (32, 33).

According to common practice in image processing, the inverse of **A** is used in applications of the spatial transform and the subsequent interpolation. To that end, in every point of the new image, the inverse transformation is applied in order to find the value that should be mapped to that grid point.

3.3 Image correspondences

The section-images will always be the starting point for the estimation of a transformation. In some cases one will be the reference for the next and from the landmarks used one can determine what kind of global transformation is estimated. This reference image is the model and therefore referred to as the model image:

> A model image is an image representing the object under study in a standard position and orientation and with standard proportions.

The image under alignment is referred to as the world image:

> A world image is an image representing the object under study in the current position and orientation and with current proportions.

Throughout the text these two images will be used. One should, however, understand that using features of proportion are allowed if and only if absolute landmarks are used (e.g. reference points).

3.4 Point-pattern matching methods

Point pattern matching is often used to find a rigid transformation; in some cases it is allowed to estimate a full affine transform. Algorithms to extract either a rigid or an affine transform from two pairs of corresponding points in the same plane will discussed. The set of corresponding points used for the estimation of an affine transform may consist of more than three points; in that case the linear system is said to be overdetermined. Collinear points do not contribute to the computation of the full affine mapping, and should be avoided. The starting point of the computation is a reference point set $P_n(M)$ and a corresponding set $P_n(W)$, transformed to $P'_n(W)$. The rationale of the estimation of geometrical mapping between two point patterns is to minimize the sum of the squared distances between point patterns $P'_n(W)$ and $P_n(M)$.

This distance is given by summing distances between corresponding point-pairs and is formulated as:

$$d = \sum_{i=1}^{i=n}[(x_i - x_i')^2 + (y_i - y_i')^2]. \qquad [5]$$

The distance d is the criterion for assessing goodness of fit of the estimated transformation as presented in *Figure 4*. The square distance is used to avoid negative values.

Estimating rigid transformation from corresponding points

To minimize d, translation and rotation have to be estimated from $P'_n(W)$ and $P_n(M)$. Since the two point-pairs are collinear, the translational difference is minimized by equating the midpoint of a line $L_m[P_1(M),P_2(M)]$ with a line $L_w[P'_1(W),P'_2(W)]$. The corresponding translation vector equals

$$\vec{t} = \left(\frac{1}{n}\sum_{i=1}^{i=n}x_i, \frac{1}{n}\sum_{i=1}^{i=n}y_i\right)_M - \left(\frac{1}{n}\sum_{i=1}^{i=n}x_i', \frac{1}{n}\sum_{i=1}^{i=n}y_i'\right)_W. \qquad [6]$$

The number of points, n in *Equation 6*, is two. Application of translation $\vec{t}$ realizes the construction of two triangles from the lines $P_1(M),P_2(M)$ and $P'_1(W),P'_2(W)$ as depicted in *Figure 5*. Rotation angle o equals the angle in point A in either of the triangles. The lengths of the sides of the triangle are calculated using points A, $P_1(M)$, and $P_1(W)'$. By rewriting the cosine rule rotation angle σ is calculated by:

$$\theta = \cos^{-1}\left(\frac{b^2 + c^2 + a^2}{2bc}\right). \qquad [7]$$

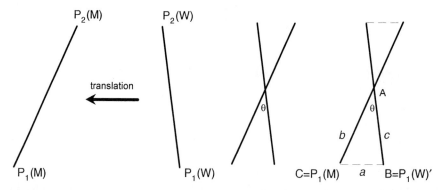

Figure 5. Estimation of rigid transformation from two points $P_1(M)$ matches with $P_1(W)$ and $P_2(M)$ matches with $P_2(W)$. After a translation over the difference vector (*Equation 6*) of the centroids of lines [$P_1(M)$, $P_2(M)$, $P_1(W)$, $P_2(W)$] a triangle is constructed so that the rotation angle can be calculated. Lengths of the triangle sides are calculated and subsequently, using *Equation 7*, the rotation angle is found.

5: Three dimensional (3D) reconstruction from serial sections

The pivot of rotation equals the midpoint of the line L_m in the case when the rigid transformation is derived from the minimal number of two corresponding point pairs. Using more corresponding points, the average of the centroid of all of the midpoints is used for the point of rotation, and a better fit might be achieved.

Estimating affine transformation from corresponding points

The determination of an affine transform from three point-pairs is straightforward. Invoking *Equation 2* it becomes clear using three corresponding point-pairs results in six simultaneous equations. The computation of the transformation matrix **A** is accomplished by:

$$(u_i\, v_i\, 1) = (x_i\, y_i\, 1) \cdot \mathbf{A}, \quad \text{where } i = 1,\, ,3. \qquad [8]$$

Vectors $(u,v,1)$ and $(x,y,1)$ are first transposed and subsequently put in an augmented form. Since, in this manner 3×3 matrices are obtained and all elements of both augmented matrices are known, an exact solution of **A** is found by solving:

$$\mathbf{A} = \begin{bmatrix} x_1 & y_1 & 1 \\ x_2 & y_2 & 1 \\ x_3 & y_3 & 1 \end{bmatrix}^{-1} \cdot \begin{bmatrix} u_1 & v_1 & 1 \\ u_2 & v_2 & 1 \\ u_3 & v_3 & 1 \end{bmatrix} \qquad [9]$$

Investigation of *Equation 5* shows that calculation of **A** requires computation of *one* inverse matrix. If more than three corresponding point-pairs are present, the system of linear equations is overdetermined and cannot be solved directly but needs to be approximated. A solution using *Equation 5* is established as the best approximation of the transformation. The approximation of this transformation is conveniently dealt with through a linear least squares approach (32) and the system is solved in a minimum square sense. To that end the points-patterns are aggregated and put in augmented matrix format; this matrix will be referred to as **X**. Its inverse cannot be directly computed. Therefore *Equation 4* is generalized to, in short:

$$\mathbf{U} = \mathbf{X}\mathbf{A}. \qquad [10]$$

In the case of a second degree approximation this amounts to:

$$\begin{bmatrix} u_1^2 \\ u_2^2 \\ u_3^2 \\ \dots \\ \dots \\ u_n^2 \end{bmatrix} = \begin{bmatrix} 1 & x_1 & y_1 & x_1 y_1 & x_1^2 & y_1^2 \\ 1 & x_2 & y_2 & x_2 y_2 & x_2^2 & y_2^2 \\ 1 & x_3 & y_3 & x_3 y_3 & x_3^2 & y_3^2 \\ \dots & \dots & \dots & \dots & \dots & \dots \\ \dots & \dots & \dots & \dots & \dots & \dots \\ 1 & x_n & y_n & x_n y_n & x_n^2 & y_n^2 \end{bmatrix} \cdot \begin{bmatrix} a_{00} \\ a_{10} \\ a_{01} \\ a_{11} \\ a_{20} \\ a_{02} \end{bmatrix}$$

In practice, finding a solution for **A** amounts to multiplication of both sides of the equation with the transposition of **X**, so:

$$\mathbf{X}^T\mathbf{U} = \mathbf{X}^T\mathbf{X}\mathbf{A}. \qquad [11]$$

Since $\mathbf{X}^T \times \mathbf{X}$ is now a square matrix, the matrix $\mathbf{A}$ can be represented as:

$$\mathbf{A} = (\mathbf{X}^T\mathbf{X})^{-1}\mathbf{X}^T\mathbf{U}. \qquad [12]$$

If $(\mathbf{X}^T \times \mathbf{X})^{-1}$ cannot be inverted a problem arises which can be solved by introducing singular value decomposition techniques (32).

Evaluation and assessment of goodness of fit
The point pattern methods need to be assessed. This is realized in two ways. The most important one is evaluation against '*d*' of *Equation 5*. By transforming the corresponding points using the estimated transformation the goodness of fit is assessed. A fit can be derived for each individual point; in this manner, outliers (mismatches) can be removed so that the fit gradually improves. If, additionally, iteration is used, the effect of inaccurate matches is diminished. This technique applies for estimation of rigid as well as affine transformation.

3.5 Shape-based assessment methods

Landmark shapes are patterns extracted from the images and matching of landmark shapes requires spatial and shape similarity. Binary images are therefore used in this matching process. Spatial correlation of silhouettes (landmarks) is determined using logical image operations. Measuring a correlation of two binary images implies establishing the extent of overlap of patterns in the images. Two measures are presented: the binary correlation and the binary overlap.

Binary correlation of image M(x,y) with image W(x,y)
The correlation function is based on the spatial distribution of pixels in the world image with respect to the model image and is expressed as a fraction of total similarity. Thus, if the patterns in the images are exactly the same the correlation is one and if the patterns do not coincide at all it is zero. The images are scanned for the number of object pixels, i.e. the 'one'-elements of the binary image.

Spatial overlap of the objects in image $M(x,y)$ and $W(x,y)$ is expressed by the number of pixels in common, given by the number of pixels resulting from a logical AND operation of images $M(x,y)$ and $W(x,y)$, written as their intersection: $N_{M(x,y) \cap W(x,y)}$. Spatial overlap is compared to the total number of pixels that could possibly contribute to the object, i.e. the total number of pixels is the image taking into account that the image sizes of the model and world images are equal. This guarantees that correlation is always in [0,1]. The correlation function $\rho(W(x,y)M(x,y))$ is written (8, 17) as:

$$\rho[M(x,y),W(x,y)] = \frac{\left[N_{M(x,y) \cap W(x,y)} \times N_{\text{tot}} - N_{M(x,y)} \times N_{W(x,y)}\right]}{\left[\left(N_{M(x,y)} \times N_{\text{tot}} - (N_{M(x,y)})^2\right) \times \left(N_{W(x,y)} \times N_{\text{tot}} - (N_{W(x,y)})^2\right)\right]^{1/2}}, \qquad [13]$$

5: Three dimensional (3D) reconstruction from serial sections

N_{tot} representing the total number of pixels in the image, and $N_{M(x,y)}$ and $N_{W(x,y)}$ representing the number of object-pixels in the model and world images respectively. It is assumed that both $N_{M(x,y)}$ and $N_{M(x,y)}$ are positive numbers.

Binary overlap of image M(x,y) with image W(x,y)

Equivalent to the correlation function, an overlap function is formulated (24) based on object-pixel patterns in a binary image and describing the extent to which the areas of the objects in $M(x,y)$ and $W(x,y)$ overlap. The area held in common by $M(x,y)$ and $W(x,y)$ is determined by an AND operation. The overlap is expressed as a fraction of total overlap; however, for computation only object pixels are taken into account, and the overlap function $\tau[M(x,y),W(x,y)]$ is written as:

$$\tau[M(x,y),W(x,y)] = \frac{2 \times N_{M(x,y) \cap W(x,y)}}{N_{M(x,y)} + N_{W(x,y)}}. \quad [14]$$

(For explanation of individual terms and assumptions, see *Box 7*.)

Functions for binary correlation are used in assessment of the goodness of fit. This assessment can be part of an iterated fine-tuning of a coarse estimation procedure. In Section 3.9 such an iterative procedure is discussed. A coarse estimation can be realized by analysing the scene using the image moments (see Sections 3.6–3.8).

Box 7. Binary cross-correlation/overlap for assessment of registration

Algorithm

Start: Get model binary image M; Get world binary image W.
sx = ImageWidth(M) = ImageWidth(W)   *for the image sizes should be equal*
sy = ImageHeight(M) = ImageHeight(W)
HI = M (**AND**) W   *outcome of the AND operation is stored in HI*
total = sx*sy.   *total number of pixels in the image*
intersect = pixel_sum (HI)   *count the number of '1' pixels in HI.*
model = pixel_sum (M)   *count the number of '1' pixels in M*
world = pixel_sum (W)   *count the number of '1' pixels in W*
nominator = (intersect*total) –   *introduce a help variable for*
   (world*model)   *nominator*
denominator = SQRT((model*total – SQR(model)*(world*total – SQR(world))
bcc = nominator/denominator   *apply Equation 13*
bo = (2*intersect)/(world + model)   *apply Equation 14*
return **bcc** and **bo**; end

3.6 Registration and congruencing using image moments

The object under study (landmark) can be described using the image moments. To that end the model object M is compared to the world object W. The objects are extracted from images $M(x,y)$ and $W(x,y)$ by a segmentation procedure. It is, therefore, assumed that the objects are represented by their silhouette as present in a binary image.

General theory of the image moments

Here, description of the image moments is restricted to the 2D case. The 2D image moments are defined in terms of an underlying continuous function (31, 34), and taking account of discrete sampling this becomes:

$$M_{pq} = \sum_{0}^{sx-1} \sum_{0}^{sy-1} x^p y^q \cdot F(x,y), \qquad [16]$$

where sx, sy denote the size of the image.

M_{pq} is the pth-order moment with respect to x and the qth-order moment with respect to y. The method of moments is described in theoretical statistics and applied mechanics. Its use in image analysis was first described by Hu (34). The method of the moments is widely used in pattern recognition problems (34–36). Here, image moments are used to resolve geometrical differences between scenes. To that end geometrical features of the distribution function are extracted from its moments. Because scenes are simplified to binary objects, the shape of the distribution function directly corresponds with the geometry of objects in the scene. Relative invariants are derived from whichever estimation of geometrical transformation is realized.

Relative invariants with respect to translation

From the set standard moments (*Equation 16*) translation invariance is achieved by centralizing the moment set about the mean of the distribution. The centralized moment set is expressed as:

$$\mu_{pq} = \sum_{0}^{sx-1} \sum_{0}^{sy-1} (x - \bar{x})^p (y - \bar{y})^q \cdot F(x,y). \qquad [17]$$

The mean of the distribution is calculated from the standard zero- and first-order moments by:

$$\bar{x} = \frac{M_{10}}{M_{00}} \quad \text{and} \quad \bar{y} = \frac{M_{01}}{M_{00}}. \qquad [18]$$

The zero-order standard moment represents the number of elements of the objects in the image, since application of *Equation 16* with $p,q = 0$ in effect sums the elements of the image that are non-zero and thus belonging to the object. The geometrical interpretation of the zero-order standard moment is the area of the objects in the image (25, 37). The arithmetic mean of the

5: Three dimensional (3D) reconstruction from serial sections

independent variables x and y is by definition the centroid of the geometrical shape. In this particular case the geometrical shape corresponds to an irradiance distribution of uniform density, therefore the centroid coincides with the centre of mass of the distribution (object).

Relative invariants with respect to orientation

The centralized moments (*Equation 17*) are used to derive two orthogonal principal axes of the object. The principal axes have an orientation with respect to the x-axis of the image. The corresponding slope angle with the x-axis is also referred to as the tilt angle. In terms of the central moments of the image, this tilt angle is given by (30, 38):

$$\varphi = \tfrac{1}{2} \tan^{-1}\left(\frac{2\mu_{11}}{\mu_{20} - \mu_{02}}\right). \qquad [19]$$

A unique tilt angle can only be determined if unique principal axes of inertia exist, i.e. the object should not be a circle.

Relative invariants with respect to proportions

If the object is considered as an ellipse, i.e. the image an ellipse—an elliptical disc with uniform intensity and the same mass as the object, then this ellipse can be defined with a semi-major and a semi-minor axis. The size of the ellipse is determined by the lengths of these axes (36, 38). Lengths of these axes are described in terms of second-order central moments. The length of the semi-major axis is written as:

$$\alpha = \left(\frac{2\cdot\left[\mu_{20} + \mu_{02} + \sqrt{(\mu_{20} - \mu_{02})^2 + 4\cdot(\mu_{11})^2}\right]}{\mu_{00}}\right)^{1/2}. \qquad [20]$$

Similarly, the length of the semi-minor axis is written as:

$$\beta = \left(\frac{2\cdot\left[\mu_{20} + \mu_{02} - \sqrt{(\mu_{20} - \mu_{02})^2 + 4\cdot(\mu_{11})^2}\right]}{\mu_{00}}\right)^{1/2}. \qquad [21]$$

The extent of the distribution function and therewith the size of the object is characterized by the lengths of the principal axes given by 2α and 2β. However, in most cases the values of α and β are used directly as relative invariants of the object.

3.7 Affine transform components estimation

The starting point of the estimation of a rigid/affine transform from the image moments are objects M and W. The estimated transformation will be used to normalize W with respect to M so that shapes W and M are congruent in a global linear sense.

Translation estimation

The centroid $C^m(x_c,y_c)$ of object M was calculated from the moments of $M(x,y)$ and likewise the centroid $C^w(x_c,y_c)$ of object W from the moments of $W(x,y)$. Consequently, position invariance of object W with respect to object M is obtained by translation over the difference vector of their centroids. The estimation of this difference vector $\vec{t}$ is expressed by:

$$\hat{t}_x = C_x^m - C_x^w = \Delta x \quad \text{and} \quad \hat{t}_y = C_y^m - C_y^w = \Delta y. \qquad [22]$$

The centroid $C^m(x_c,y_c)$ will also be used as a pivot of transformation with respect to rotation and scaling.

Rotation estimation

From the second-order moments the angle with respect to the x-axis of the object M, θ^m, is calculated. Likewise, the tilt angle for object W, θ^w, is obtained. The rotation angle bringing W into register with M, normalizing orientation of W with respect to M, is estimated by:

$$\hat{\theta} = \hat{\theta}^m - \hat{\theta}^w. \qquad [23]$$

The angle θ^m is required for orientational normalization of the object axes with the x- and y-axis as a part of the scaling transformation, that is if scaling is applicable.

Scaling estimation

From the second-order moments the lengths of the semi-major and the semi-minor axes were calculated. The ratio of these lengths equals the scaling necessary to make object W proportional to object M, i.e. normalization of object W with respect to the proportions of object M. Scaling estimation is denoted by the scaling vector $S(S_x, S_y)$. The components of $S(S_x, S_y)$ are respectively given by:

$$\hat{S}_x = \frac{\alpha^m}{\alpha^w} \quad \text{and} \quad \hat{S}_y = \frac{\beta^m}{\beta^w}. \qquad [24]$$

Both components of the vector are correctly estimated under the assumption that the major-principal axis is established as the axis that intersects the x-axis with the angle φ of *Equation 19*.

3.7.1 Concatenation of estimations to one affine transformation

Application of the estimation of a translation, rotation, and scaling from the image moments is now concatenated into one transformation. Proper application of the spatial transformation requires the inverse of the concatenated transformation (see Section 3.2). The inverses of the respective separate

5: Three dimensional (3D) reconstruction from serial sections

transformations are used, instead of first concatenating the transformation and henceforth computing the inverse of the concatenated transformation matrix. This could be realized as the components of all individual transformations are known. Each of the individual transformations is therefore written as its inverse in homogeneous matrix notation. Thus:

$$T^{-1} = \begin{pmatrix} 1 & 0 & 0 \\ 0 & 1 & 0 \\ -\Delta x & -\Delta y & 1 \end{pmatrix}. \quad [25]$$

the inverse translation over the estimated translation vector $\vec{t}$ (*Equation 21*), followed by,

$$R^{-1} = \begin{pmatrix} -1 & -0 & 0 \\ -0 & -1 & 0 \\ -C_x^M & -C_y^M & 1 \end{pmatrix} \cdot \begin{pmatrix} \cos(-\hat{\theta}) & \sin(-\hat{\theta}) & 0 \\ -\sin(-\hat{\theta}) & \cos(-\hat{\theta}) & 0 \\ -0 & 0 & 1 \end{pmatrix} \cdot \begin{pmatrix} 1 & 0 & 0 \\ 0 & 1 & 0 \\ C_x^M & C_y^M & 1 \end{pmatrix}, \quad [26]$$

the inverse rotation over the estimated rotation angle. Since the object in M(x,y) is considered the model, the centroid of M is used as the pivot for rotation. This requires a back translation to the origin to be included in the transformation. The inverse rotation is followed by:

$$S^{-1} = \begin{pmatrix} -1 & -0 & 0 \\ -0 & -1 & 0 \\ -C_x^M & -C_y^M & 1 \end{pmatrix} \cdot \begin{pmatrix} \cos(-\theta^M) & \sin(-\theta^M) & 0 \\ -\sin(-\theta^M) & \cos(-\theta^M) & 0 \\ -0 & 0 & 1 \end{pmatrix} \cdot \begin{pmatrix} (S_x)^{-1} & 0 & 0 \\ 0 & (S_y)^{-1} & 0 \\ 0 & 0 & 1 \end{pmatrix} \cdot$$

$$R^{-1} = \begin{pmatrix} -\cos(-\theta^M) & \sin(-\theta^M) & 0 \\ -\sin(-\hat{\theta}^M) & \cos(-\hat{\theta}^M) & 0 \\ 0 & 0 & 1 \end{pmatrix} \cdot \begin{pmatrix} 1 & 0 & 0 \\ 0 & 1 & 0 \\ C_x^M & C_y^M & 1 \end{pmatrix}, \quad [27]$$

the inverse scaling by scaling vector S. To assure proper scaling, first alignment of the major axis with the x-axis is realized by back rotation over the angle of inclination of object M. Back rotation should be preceded by back translation to the origin since the centroid of M is used as pivot for rotation. Consequently, concatenation of the individual transformations into one transformation is written as:

$$A^{-1} = T^{-1} \cdot R^{-1} \cdot S^{-1}, \quad [28]$$

the inverse of the transformation matrix representing the global congruencing transformation. This transformation can now be applied to the image that W(x,y) (binary image) was derived from. Note that the centroid and angle of inclination of object M are used to derive A^{-1}. If only rigid transformation is allowed, then *Equation 27* is not used and *Equation 28* is a concatenation of *Equations 25* and *26*.

> **Box 8.** Resolving the ambiguity in the rotation estimate
>
> *Algorithm*
>
> The angle φ is computed from *Equation 19*. The angle θ is used as in *Equation 23*. This choice is made because from *Equation 19* it cannot be unambiguously established that φ is the angle of the major or the minor axis with the *x*-axis. From the second-order centralized moments two features are computed: the major variance, $mv = \mu_{20} - \mu_{02}$, and Pearson's correlation coefficient,
>
> $$pcc = \frac{\mu_{11}}{\sqrt{(\mu_{20} \cdot \mu_{20})}} \quad (8).$$
>
> If $mv > 0$; the major variance is the largest along the *x*-axis.
>   if ($pcc > 0$) then: $\theta = \varphi$
>   else: $\theta = -(\varphi_0 - \frac{1}{2}\pi)$
> If $mv < 0$; the major variance is the largest along the *y*-axis.
>   if $pcc > 0$ then: $\theta = \frac{1}{2}\pi - |\varphi_0|$
>   else: $\theta = -(\frac{1}{2}\pi - \varphi_0)$
> Application of this algorithm reveals the ambiguity and the correct angle of θ is used in *Equation 23*.

3.8 Evaluation using similarity measures

The transformation estimated from the image moments requires assessment of goodness of fit and to this end *Equations 13* and *14* are used. Application of the estimated transformation on $W(x,y)$ should result in high values for binary correlation and binary overlap. This evaluation is required as the moments are only very global and tend to smooth out small differences. Moreover, noise on the edge of the silhouette is amplified by the image moments resulting in a less good fit.

The functions of *Equations 13* and *14* can be used in a fine-tuning operation. After application of the estimated transformation a number of translations and rotations are probed in a restricted neighbourhood near the current position and orientation of the object. The position and/or orientation with maximum binary correlation is chosen as the best fit and the transformation is adapted accordingly. In effect this method results in the estimation of a rigid transformation. The overall rigid transformation that is applied after N iterations of rotation and translation optimization can be obtained by concatenation of the transformations resulting from each iteration, i.e. written in homogeneous coordinate format:

$$\mathbf{A} = \prod_{n=0}^{N} \left[\begin{pmatrix} -\cos\varphi_n & \sin\varphi_n & 0 \\ -\sin\varphi_n & \cos\varphi_n & 0 \\ -rx & ry & 1 \end{pmatrix} \cdot \begin{pmatrix} 1 & 0 & 0 \\ 0 & 1 & 0 \\ \Delta x_n & \Delta y_n & 1 \end{pmatrix} \right]_n, \quad [29]$$

where $rx = [(1 - \cos \varphi_n) \cdot c_x + \sin \varphi_n \cdot c_y]$, $ry = [(1 - \cos \varphi_n) \cdot c_y + \sin \varphi_n \cdot c_x]$, and (c_x, c_y) denotes the point of rotation for which in practice the centroid of the object in the model image is used.

The complexity of the estimation of the congruencing transformation can be expanded by invoking the scaling in a similar fashion as rotation and translation. However, as scaling is applied preferably along the object axes, methods that are applied from the landmark shapes rather than the landmark frames are intrinsically better equipped for this application.

3.9 Intensity-based methods

In the previous chapter binary cross-correlation was used to optimize an estimated transformation. The use of landmark frames (see Section 2.5.2) is based on evaluation of intensity values and therefore other methods are required. A landmark frame of the model is matched to a landmark frame of which the position and orientation are to be put into correspondence with the model image. Based on the cross-correlation of the two images a correction for position and orientation is found. Cross-correlation can be dealt with in either the spatial domain or the frequency domain. According to a theorem it is stated that the correlation calculated in the spatial domain is equivalent to multiplication in the frequency domain (31). The cross-correlation is expressed in the energy spectrum which is obtained by a multiplication of the Fourier spectrum of the model image with the complex conjugate of the Fourier spectrum of the world image and then taking the inverse Fourier transform.

Estimation of translation by cross-correlation

A shift of the model image with respect to the world image is expressed by a shift of the maximum of the cross-correlation. If the model is correlated with itself, referred to as autocorrelation, the correlation maximum is located at the centre of the magnitude image. If the model image is shifted by a vector $T(\Delta x, \Delta y)$ then the maximum of the correlation function is also shifted by $T(\Delta x, \Delta y)$. Examination of the position of the maximum of the correlation function yields a translation imposed earlier. In this way translation of the world image with respect to the model image is estimated.

Estimation of rotation by cross-correlation

To establish a possible rotation of the world image with respect to the model image, an extra transformation is necessary because no direct relation of a shift of the maximum of the correlation function with respect to a rotation exists. To that end the image is transformed to a polar coordinate system (φ, r), where φ represents the polar angle and r the polar radius (24). The cross-correlation in the polar coordinate system is used to find the rotation angle since it relates to the Cartesian coordinate system. Namely, translation in polar coordinates equals rotation in the Cartesian coordinate system.

Consequently, shift of the maximum of the correlation function of Δφ in the polar coordinate system corresponds to rotation of Δφ in the Cartesian coordinate system. In this way rotation of a world image with respect to a model image is established.

Protocol 3.   Image registration

Materials
- Get model image: M
- Get world image: W

Translation method
1. Apply fast Fourier transform to both images: FFT(W), FFT(M)
2. Compute the complex conjugate to get the magnitude of the Fourier spectrum: FFT(W)*FFT(M).
3. Compute the inverse FFT of the complex conjugate: $(FFT(W)*FFT(M))^{-1}$.
4. Determine the locus of the maximum.
5. The shift from the centre (centrepoint of the image) represents the translation vector.

Rotation method
1. Resample (counterclockwise) the images from the centrepoint in new images $M(\varphi,r)$, $W(\varphi,r)$.
2. Continue as for translation.
3. The shift should from the centre along the φ-axis is equivalent to the rotation in radians.

---

4. Systems for routine application of 3D reconstruction

Several methods have been designed to work directly with greyvalue images. Two of these systems depending on dedicated hardware are described in this section. A general principle with image fiducials is to obtain a coarse fit using additional hardware tools and to refine this fit by software matching procedures. Hardware is used to reduce the number of matches to evaluate to an acceptable number. In most cases a fitting procedure for image frames will consist in finding the best possible registration expressed by a rigid transformation. Two examples of landmark frames will be discussed. The first method is based on comparison of images of subsequent sections, the second introduces an extra image that contains information on the object before sectioning.

4.1 3D reconstruction under rigid transformation

In classical reconstruction methods a technique, known as the optical shuttle method (3, 39), was developed using a discussion microscope (combined optical path of identical microscopes) to manually match and register the current image to a previous one. With advanced image processing hardware becoming available, comparable systems have been developed using the overlay of two consecutive section-images in a video buffer (40). In these systems (initial) registration is performed manually. The result of the manual registration is comparable to a result obtained using cross-correlation. The advantage of the manual method is the decisions made by the expert user in the case of imperfection in sections.

4.1.1 Mode of operation and practical implementation

With modern technology a system for manual alignment and semi-automated evaluation has been developed. In *Figure 2C* a microscope is depicted which is equipped with a motorized *xy*-stage and a camera attached to the microscope by a rotatable mounting stage. Three directions can be adjusted by the user, the *x*- and *y*-directions and, a rotation angle. Adjustments are accomplished by a joystick attached to the steppermotor controller. The framegrabber employed to capture images is capable of keeping one image in memory and applying a real-time subtraction by a look-up table (LUT) operation. The image of the last section is kept in memory whereas the image of the current section is captured, the difference is obtained in real-time and can be adjusted until the difference is minimal, or a best fit according to the specialized end-user. An image of a scene matched with itself would reveal no differences. In *Figure 6* a result of such a manual fit is given. To convey the idea red and blue LUTs were used instead of quenching grey LUTs.

4.1.2 Maximization of cross-correlation for fine-tuning

Once a reasonable fit is realized, the section-image is captured in the standard operating modus of the framegrabber. Sequential application of this technique results in a practically well-aligned stack of images. At this point the computer is employed to fine-tune the alignment using objective criteria in a restricted search space, using maximization of cross-correlation as measure for assessment. A necessary tool in the acquisition of images with this system is a good administration of sequential image acquisition (see Section 2.3.1). In *Figure 7*, an example obtained by using this system is depicted.

4.2 3D Reconstruction including deformation correction

A method including a possible full geometrical restoration has the following features:

- Deformation introduced by the histological processing must be measurable and correctable.
- Procedures for registration should rely on absolute structures of reference.

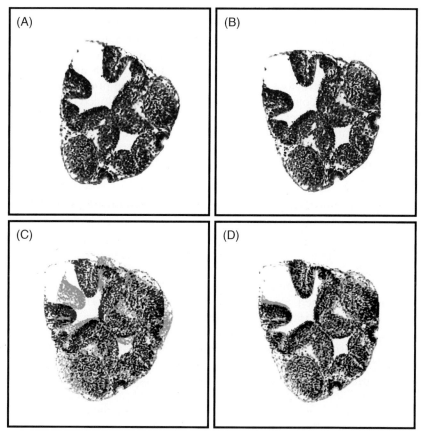

Figure 6. Example of alignment using grey-value images. For illustrations, two subsequent section images of a 48-h zebrafish embryo were used. One of these (B) is taken as reference for the position and orientation and the other image (A) is registered to that position and orientation. Images are acquired with a CCD camera (Sony XC-77CE) mounted on a Zeiss Axioskop using a 5× objective. The sectioning was performed obliquely to the anterior–posterior (AP) axis of the embryo. (**A**) An image of a section in its native orientation and position. (**B**) The model image. (**C**) Superposition of images A and B using the *xy*-stage of the microscope. (**D**) Superposition of images A and B using the *xy*-stage of the microscope, camera rotation, and fine-tuning by cross-correlation. The image of (A) is now registered with respect to the position and orientation of the image in (B).

A system based on these requirements may mimic some of the advantages normally attributed to non-invasive sectioning techniques, namely:

- An alignment between adjacent sections over a whole stack of image-section-images.
- Absence of geometrical distortions in the *xy*-plane.

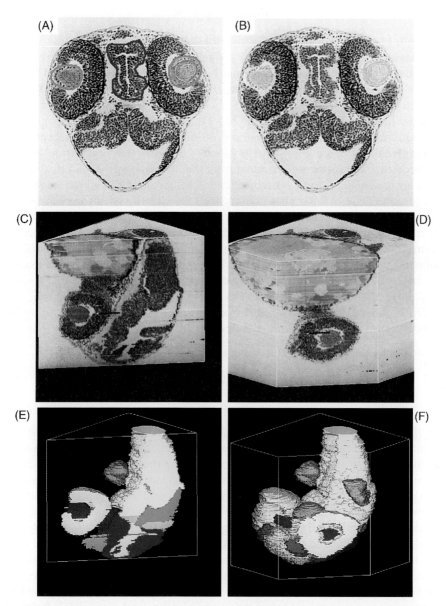

Figure 7. Illustration of reconstruction rendering techniques using the data of a 48-h pf. zebrafish embryo (head region), sectioned almost perpendicular to the AP axis. (A) Annotation of structures in a section-image (zebrafish 48-h pf.) using the contour representation. In (B) the same structures are indicated with voxel planes that are transparently superimposed over the grey-value image. In (C, D) a volume representation of the stack of section-images is rendered in a new plane of sectioning using an electronic knife, obtained by the method described in section 4.1. In (E,F) volume renderings of the annotated counterparts of the image stack, using a ray-casting visualization technique, are shown. The same part as in (C, D) is cut away. Absence of background alows visualization the whole head region (compare C with E and D with F). All relevant structures in the zebrafish head were annotated.

System for application of deformation correction

A system is accomplished by the introduction of an extra image, an absolute landmark frame, that is captured just before a section is cut (8, 9, 17). An image taken from the block containing the object would most certainly be free of deformation due to sectioning. Since complete alignment of subsequent images has to be achieved, it should be guaranteed that the images are captured in the same orientation and position. In this set-up, image capture is always realized from the same position, and also the orientation of the embedding block does not change during the sectioning process. For this set-up a horizontal microtome is used, extended with a dedicated image acquisition microscope (*Figure 2D*). The sections are subsequently used to probe for internal structures and section-images are acquired with a conventional research microscope. Each section-image is attributed with an absolute reference in terms of change in position and orientation, as well as possible deformation inflicted to the section. The resulting image pair is conveniently arranged in an *image set* consisting of a model and a world image (see Section 3.3). Consequently, exact registration as well as deformation correction is possible.

Differences in images can result from the processing technique or imaging conditions, and these should be taken into consideration, that is:

- the object is subject to a processing technique
- contrast enhancement of a specific part of the object (e.g. staining, bandpass filter)
- the use of a different lens–sensor system

In the image-set the model image is always used for registration and deformation correction of the other image in the set. The model image is a view of the surface of the embedding block containing the object under study prior to sectioning imaged with episcopic illumination, and consequently the model image is referred to as an episcopic image. The world image is a section-image submitted to histochemical procedures. This image is obtained with a microscope by transmitted light, i.e. diascopic illumination, and consequently the world image is referred to as a diascopic image.

The microtome is adapted so that the episcopic image is always taken at the same position and in the same orientation. Therefore, all episcopic images are already in register. Since in these images the object is not yet exposed to deformation, proportional information can also be used as a model. Relating the diascopic images to the episcopic images allows one to produce a 3D reconstruction with the advantages of both images.

Features of the episcopic-model and the diascopic-world image

The episcopic image of the object under study has the following characteristics:

- view of the surface of the object by a microscope with episcopic illumination

5: Three dimensional (3D) reconstruction from serial sections

- model of the section-image with respect to position, orientation, and proportion
- no addition of specific contrast to the object

The diascopic image of the object under study has the following characteristics:

- section image of the object by a microscope with diascopic illumination
- reflects current position, orientation, and proportion of the section
- contains specific contrast, added to the object by a histochemical staining procedure

Examples of episcopic and diascopic images are shown in *Figure 8*. The principal difference between the episcopic and the diascopic images is that the episcopic image contains only shape information on the object whereas the diascopic image may also contain information on (gene-expression) patterns in the object depending on the staining applied.

4.2.1 Normalization to a model by congruencing

The model (episcopic) image provides the archetype of shape. The distortion inflicted on the object is represented by the world image. Through the combination of information extracted from these images an undistorted 3D reconstruction (3D image) can be obtained. To that end the diascopic image is normalized with respect to position, orientation, and proportion, taking the episcopic image (the model) as norm: *congruencing normalization*. The normalization procedure ideally consists of just one congruencing operation. However, as pointed out in Section 2, the section might have been exposed to non-linear deformations and therefore one global congruencing operation will not always be sufficient for geometrical restoration, and local congruencing is required for complete restoration. Congruencing normalization using this system consists of the following:

- Select and extract landmark from $M(x,y)$ and $W(x,y)$.
- Apply global congruencing transformation according to estimated transformation.
- Assess the goodness of fit of the global transformation.
- Look for local differences.
- Apply local congruencing transformation.
- Assess the goodness of fit after application of the local congruencing transformation.

This sequence of operations corresponds with *Figure 4*, now local transformations are also used. It is assumed that global differences can be restored using a global congruencing transformation. If the images already nearly fit, the last differences are resolved by locally operating transformations. These

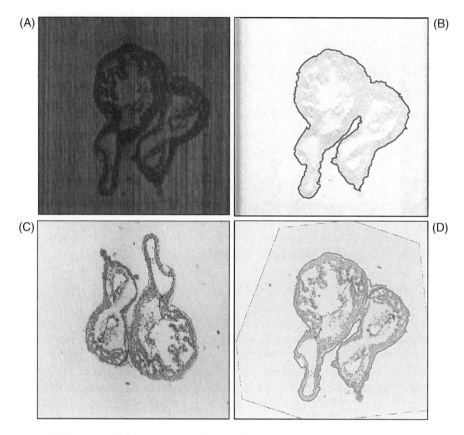

Figure 8. Example of alignment according to the episcopic–diascopic set-up (see section 4.2) and illustrated with images of an isolated 11 Embryonic Days (ED) rat heart; itself is used as the landmark shape. The congruencing normalization is estimated on the basis of binary images obtained by segmentation. The contour of the episcopic image is superimposed to give a visual clue of the goodness of fit. Images are captured with a Videk high-resolution CCD camera mounted on a microscope (4× objective lens). (A) Episcopic image of an 11 ED rat heart. The structure of the surface of the embedding block is clearly visible by the vertical burrs. These burrs complicate the segmentation and therefore a restoration technique was applied (8). (B) Episcopic image of (A) after restoration. The contour that was obtained from the segmentation is superimposed in magenta. (C) Unregistered diascopic image corresponding to (A, B). Good resolution of the histology, typical for the diascopic image, is clearly visible. (D) Result of the registration–deformation correction of the diascopic image of (C) with respect to the episcopic image. The contour of the episcopic image is again superimposed in cyan to appreciate the goodness of fit.

5: Three dimensional (3D) reconstruction from serial sections

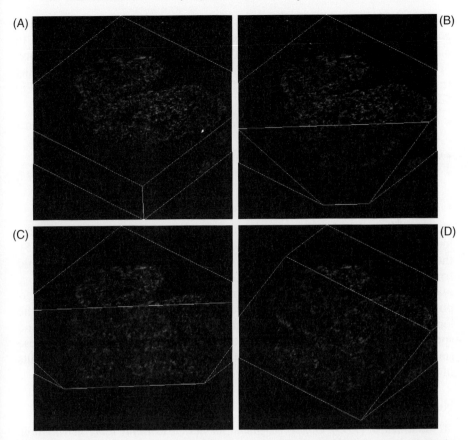

Figure 9. A visualization of a part (22 sections) of a 3D reconstruction of an embryonic liverlobe (16 ED rat). The areas with bright texture indicate the presence of the liver specific enzyme GDH, probing cells that have differentiated into hepatocytes. The accuracy of the reconstruction can be appreciated by looking at the alignment around the intersections of the major blood-vessels in the liverlobe. In B–D an oblique cutting plane is introduced. Moving with this cutting plane through the 3D image allows the image alone a new plane of sampling to be seen, thus confirming the same quality of alignment. For appreciation of the method, pay particular attention to the geometrical fit around the intersections of blood-vessels. Moreover, random distribution of hepatocytes is observed on passing through the object other than in the direction of sampling.

local transformations are affine transformations which are applied in a restricted region of the image.

4.2.2 Matching of images from the image set

Although images have a 1:1 relation, the intensities of the images cannot be compared directly. Reference points and landmark shapes are good candidates to use in the matching operation for estimation of parameters for congruencing normalization. Landmark shapes will allow easy assessment of goodness of fit. For extraction of the landmark shapes a segmentation to binary images is therefore required. In *Figure 8* an example of the application of extraction of landmark shapes and subsequent application of congruencing transformation is depicted. The transformation is estimated as described in Sections 3.6–3.8:

- Compute the global congruencing transformation from the image moments (*Equation 28*).
- Assess the goodness of fit after transformation (*Equations 13* and *14*).
- Fine tuning of global transformation might be necessary using *Equations 13, 14,* and *29*.
- Assess the goodness of fit and look for local differences.

Now that the world image is made almost congruent with the model image, a local assessment is required. From the intersection image (see Section 3.5), the local differences are easily found. The difference areas are divided into triangular regions. Each of the triangular regions in the world image is mapped to the model image using a three-point affine transformation. Interpolation of these areas is realized by scan conversion (33). In some cases the local transformation need to be smoothed over the entire image. To this end special smoothing functions can be applied (8, 17). Application of these smoothing functions to local transformations realizes a non-linear fit. Results of a reconstruction using this system are depicted in *Figure 9*.

References

1. Buurman, J. (1993). *Object recognition for flexible assembly using stereo vision.* PhD thesis, Delft University Press.
2. Ware, R. A., LoPresti, V. (1975). *International Review of Cytology,* **40**, 325–440.
3. Gaunt, W. A. and Gaunt, P. N. (1978). *Three dimensional reconstruction in biology.* Pitman Medical Publishing Co. Ltd, Tunbridge Wells.
4. van der Voort, H. T. M., Brakenhof, G. J., Valkenburg, J. A. C., and Nanninga, N. (1985). *Scanning,* **7**, 66–78.
5. Rydmark, M., Jansson, T., Berthold, C. H., and Gustavsson, T. (1992). *J. Microscopy,* **165**, 29–47.
6. Verbeek, F. J., de Groot, M. M., Huijsmans, D. P., Lamers, W. H., and Young, I. T. (1993). *Computerized Medical Imaging and Graphics,* **17**, 151–163.
7. Verbeek, F. J., Huijsmans, D. P., Baeten, R. W. A. M., Schoutsen, C. M., and Lamers, W. H. (1995). *Microscopy Research and Technique,* **30**, 496–512.
8. Verbeek, F. J. (1995). *Three dimensional reconstruction from serial sections including deformation correction.* PhD thesis, Delft University of Technology.
9. Verbeek, F. J. (1996). *Microscopy and Analysis (UK),* **Nov.**, **56**, 33–35.
10. Baldock, R. A., Verbeek, F. J., and Vonesh, J.-L. (1997). *Seminars in Cell and Developmental Biology,* **8**, 409–507.
11. Verbeek, F. J. and Huijsmans, D. P. (1998). In *Databases in biomedical research* (ed. S. T. C. Wong), pp. 117–144. Kluwer Academic, Boston.

5: Three dimensional (3D) reconstruction from serial sections

12. Kriete, A. and Pepping, T. (1992). In *Visualisation in biomedical microscopies*, pp. 329–359. VCH, Weinheim.
13. Jacobs, R. E. and Fraser S. E. (1994). *Science*, **263**, 681–684.
14. Huijsmans, D. P., Lamers, W. H., Los, J. A., and Strackee, J. (1986). *The Anatomical Record*, **216**, 449–470.
15. McLean, M. R. and Prothero, J. (1991). *Analytical and Quantitative Cytology and Histology*, **13**, 269–278.
16. Moss, V. A. (1992). In *Visualisation in biomedical microscopies, 3D imaging and computer applications*, pp. 19–44. VCH, Weinheim.
17. Verbeek, F. J. (1992). *Proceedings 11th IAPR International Conference on Pattern Recognition, Vol. III, Conference C: Image, speech, and signal analysis (ICPR11, The Hague, The Netherlands)*, pp. 347–350. IEEE Computer Society Press, Los Alamitos, California.
18. Russ, J. C. (1986). *Practical stereology*. Plenum Press, New York.
19. Inoue, S. and Spring, K., (1997). *Video microscopy*, Plenum Press, New York.
20. Dean, P., Mascio, L., Ow, D., Sudar, D., and Mullikin, J. M. (1990). *Cytometry*, **11**, 561–569.
21. Haralick, R. M. and Shapiro, L. G. (1991). *Pattern Recognition* **24**, 69–93.
22. Goshtasby, A. (1988). *Image and Vision Computing*, **6**, 255–261.
23. Hibbard, L. S., McGlone, J. S., Davis, D. W., and Hawkins R. A. (1987). *Science*, **236**, 1641–1646.
24. Hibbard, L. S. and Hawkins, R. A. (1988). *J. Neuroscience Methods*, **26**, 55–74.
25. Young, I. T. (1988). *Analytical and Quantitative Cytology and Histology*, **10**, 269–275.
26. Merikel, M. (1988). *Computer Graphics and Image Processing*, **42**, 206–219.
27. Gerrard, H. M., Andre, J. C., and Mallet, J. L. (1993). *European Microscopy and Analysis*, **Jan.**, **21**, 23–25.
28. Olivo, J.-C., Izpisúa-Belmonte, J.-C., Tickle, C., Boulin, C., and Duboule, D. (1993). *BioImaging*, **1**, 151–158.
29. Brändle, K. (1989). *Computers and Biomedical Research*, **22**, 52–62.
30. Rosenfeld, A. and Kak, A. C. (1982). *Digital picture processing*, Vol. 2. Academic Press, Orlando.
31. Gonzalez, R. C. and Woods, R. E. (1992). *Digital image processing*. Addison-Wesley, Reading, MA.
32. Press, W. H., Flannery, B. P., Teukolsky, S. A., and Vetterling, W. T. (1988). *Numerical recipes in C*. Cambridge University Press.
33. Wolberg, G. (1990). *Digital image warping*. IEEE Computer Society Press, Los Alamitos, CA.
34. Hu, M. K. (1962). *IRE Transactions on Information Theory*, **IT8**, 179–187.
35. Sadjadi, F. A. and Hall, E. L. (1978). *Proc. IEEE Conf. Pattern Recognition and Image Processing, Chicago*, **II**, 181–187.
36. Prokop, R. J. and Reeves, A. P. (1992). *Computer Vision Graphics and Image Processing: Graphical Models and Image Processing*, **54**, 438–460
37. Young, I. T. (1995). *IEEE Engineering in Medicine and Biology*, **15**, 59–66.
38. Teague, M. R. (1979). *J. Opt Soc. Am.*, **70**, 920–930.
39. Zimmermann, M. H. and Tomlinson, P. B. (1966). *Science*, **152**, 72–73.
40. Mark, M., Lufkin, T., Vonesh, J. L., Ruberte, E., Olivo, J. C., Dolle, P., Gorry, P., Lumsden, A., and Chambon, P. (1993). *Development*, **119**, 319–338.

6

3D analysis: registration of biomedical images

DANIEL RUECKERT and DAVID J. HAWKES

1. Introduction

The analysis of medical images plays an increasingly important role in many clinical applications. Different imaging modalities provide anatomical information about the underlying tissues such as the X-ray attenuation coefficient from X-ray computed tomography (CT) and proton density or proton relaxation times from magnetic resonance (MR) imaging. The images allow clinicians to gather information about the size, shape, and spatial relationship between anatomical structures and any pathology, if present. Other imaging modalities provide functional information such as the blood flow or glucose metabolism from positron emission tomography (PET) or single-photon emission tomography (SPECT), and permit clinicians to study the relationship between anatomy and physiology. Finally, histological images provide another important source of information which depicts structures at a microscopic level of resolution.

In recent years, there has been a rapidly growing demand for the automated combination of these different sources of information. The fusion of these images is important in a number of applications such as diagnosis, therapy, and monitoring of disease. However, this integration requires a *registration* step which maps the images in a common space. Recent breakthroughs have led to the development of very robust and accurate registration algorithms. The purpose of this chapter is to give a comprehensive overview of existing techniques for medical image registration. We will also give a step-by-step description of the design of registration algorithms and their basic components for specific applications.

The majority of registration algorithms can be classified according to their application area. Therefore we will first give a brief overview of these application areas. A more detailed classification and review of registration techniques can be found in (1–3).

1.1 Intra-subject registration

The different medical imaging modalities can be used to provide spatial measurements of different (and often complementary) anatomical or functional properties of the same subject. One such example is the registration of MR and CT images of the head for surgery and therapy planning (4): on the one hand, CT provides good contrast between soft tissue and bone but is not well suited for the differentiation of different soft tissues. On the other hand, MR is not well suited for locating bony structures but provides very good contrast between many soft tissues. Other examples include the registration of MR and SPECT/PET images of the brain to localize tracer uptake anatomically to indicate brain physiology (5), or of the head and neck for cancer detection and staging (6). These applications are termed intra-subject registration since the images are acquired from the same subject and therefore represent different views of the same structure.

1.2 Inter-subject registration

Inter-subject registration deals with the problem of registering images across different subjects. For example, the registration of MR images is required to assess the morphometric variability of structures over a large number of individuals (7). In other cases, the registration of images from different subjects can improve the statistical significance of the findings, for example in functional images such as PET (8). A closely related problem is the registration of an atlas to images of individuals in order to compare the size and shape of structures among and across groups. Atlas registration is also used for the model-based identification and delineation of anatomical structures, particularly in the brain and in areas where the identification of boundaries between structures is difficult due to little or no image contrast (9). In general, the atlas itself is constructed by averaging a set of registered images of different individuals which are mapped into a common space such as the Talairach space (10). Similar problems arise for the alignment of histological images and MR for detailed studies of the human brain (11).

1.3 Serial registration

Another important role of medical imaging is the monitoring of changes in shape, size, and function of anatomical structures and pathology in individuals over short or long periods of time. Computer assisted image registration of serial images in combination with careful quality control of imaging devices provides the means for monitoring changes and assessing treatment much more sensitively than by visual interpretation alone (12–16).

1.4 Image to physical space registration

In the recent past there have been significant advances and developments in using images for navigation in certain types of surgery, particularly in the brain, skull base, the maxillofacial region, temporal bone, and spine (17, 18). Here the goal is the registration of pre-operative images, defined in a coordinate system related to the original scanning devices, to the physical space of the patient lying on the operating table. This registration allows the surgeon to use the pre-operative data for guidance. The correspondence can be provided by *extrinsic* devices, such as marker pins or stereotactic frames, which are fixed to the patient's skull during both scanning and the operation. Recently, frameless image guided surgery has been introduced, in which *intrinsic* features in the form of anatomical landmarks are used for registration. Alternatively, intra-operative images such as video, ultrasound, or X-ray imaging can be acquired. These images are therefore used primarily to provide spatial localization via registration to the pre-operative data rather than for diagnostic purposes. The same principles apply to images used to plan and guide other therapies, in particular radiotherapy. Even though image to physical space registration techniques are important for a number of applications, we will focus in this chapter primarily on image to image registration techniques. A detailed overview of image to physical space registration techniques is given by Lavallée (19).

1.5 Overview

Any registration procedure can be separated into three basic components. First, the *registration transformation* which defines the spatial relationship between both images. Second, the *registration basis* which characterizes the type of features used to establish a correspondence between both images. Finally, the *optimization*, which is used to calculate the optimal transformation parameters, forms an essential part of the registration procedure. An overview of the protocol for a generic registration procedure is shown in *Protocol 1*. All three components are closely linked to the specific application of the registration procedure. In the following sections, we will discuss these different components in more detail.

Protocol 1. A generic registration procedure

1. Define a *registration transformation* to represent the spatial relationship between both images. The properties of the transformation should depend on the nature of the deformation which is expected:

    (a) Rigid or affine transformations are suitable for intra-subject registration tasks with little physical deformation or motion, i.e. images of the head, spine, or pelvis.

    (b) Elastic or fluid transformations are suitable for inter-subject registration tasks which have to accommodate anatomical variability across subjects or soft-tissues that deform over time.

Protocol 1. Continued

2. Define a *registration basis*. The nature of the registration basis describes the types of features used during the registration process:
   (a) Point-based features such as extrinsic landmarks (like skin or bone implanted markers) and intrinsic landmarks (like anatomical or geometrical landmarks).
   (b) Contour or surface-based features of anatomical structures.
   (c) Voxel-based features including intensities and other derived features such as edges and texture.
3. Define an *optimization* procedure to calculate the optimal transformation parameters. In many cases it is not possible to calculate the optimal parameters directly from the available features. Instead an iterative optimization must be used.

2. Registration transformation

The nature of the transformation which relates two images depends on the imaging modality, the structure imaged, and the time between both image acquisitions. The flexibility of the transformation is primarily determined by the number of degrees of freedom which control the deformation. We can distinguish between rigid, affine, projective, and elastic or fluid transformations.

2.1 Rigid transformation

A common assumption in medical image registration is that both images are related by a rigid transformation. For example, for images of the head the rigid-body assumption is normally justified as the skull is rigid and constrains the motion of the brain sufficiently. In 3D, a rigid transformation involves six degrees of freedom: three rotations and three translations. Using homogeneous coordinates (20), the rigid transformation can be expressed in matrix form as

$$\mathbf{T}_{\text{rigid}}(x,y,z) = \begin{pmatrix} x' \\ y' \\ z' \\ 1 \end{pmatrix} = \begin{pmatrix} r_{11} & r_{12} & r_{13} & t_x \\ r_{21} & r_{22} & r_{23} & t_y \\ r_{31} & r_{32} & r_{33} & t_z \\ 0 & 0 & 0 & 1 \end{pmatrix} \begin{pmatrix} x \\ y \\ z \\ 1 \end{pmatrix} \qquad [1]$$

so that t_x, t_y, and t_z define the translations along the axes of the coordinate system while the coefficients r_{ij} are the result of the multiplication of three separate rotation matrices which determine the rotations about each coordinate axis.

2.2 Affine transformation

In many cases the dimensions of the voxel cannot be determined reliably. In particular, the estimate of voxel dimensions in MR imaging systems is accurate only to within a few percent. Therefore, it may be necessary to recover not only the rigid transformation parameters, but also additional scaling parameters (21). This additional scaling can be expressed in matrix form as

$$\mathbf{T}_{scale} = \begin{pmatrix} s_x & 0 & 0 & 0 \\ 0 & s_y & 0 & 0 \\ 0 & 0 & s_z & 0 \\ 0 & 0 & 0 & 1 \end{pmatrix} \qquad [2]$$

where s_x, s_y, and s_z define the scaling along the different coordinate axes. In some cases it may also be necessary to correct for shears, for example caused by the gantry tilt of CT scanners. A shear in the x–y plane can be expressed as:

$$\mathbf{T}_{shear}^{xy} = \begin{pmatrix} 1 & 0 & sh_x & 0 \\ 0 & 1 & sh_y & 0 \\ 0 & 0 & 1 & 0 \\ 0 & 0 & 0 & 1 \end{pmatrix} \qquad [3]$$

Combining the rigid transformation matrix with the scaling and shearing matrices yields an affine transformation

$$\mathbf{T}_{affine}(x,y,z) = \mathbf{T}_{shear} \cdot \mathbf{T}_{scale} \cdot \mathbf{T}_{rigid} \cdot (x,y,z,1)^T \qquad [4]$$

whose 12 degrees of freedom represent rotations, translations, scaling, and shears. Like rigid transformations, affine transformations are global in the sense that they affect the entire image domain. In a similar fashion higher-order global transformations such as trilinear (24 degrees of freedom) or quadratic (30 degrees of freedom) transformations can be modelled (22).

2.3 Projective transformation

Projective transformations play an important role in applications which involve the alignment of 3D volumes like MR or CT datasets to 2D images such as X-ray and video images. Different types of projections including parallel or perspective projections can be used depending on the application (20). However, in most cases the transformation which relates the 3D and 2D images is a combination of a projective transformation as well as a rigid transformation which determines the pose of the 3D volume relative to the camera. Often it is possible to determine the perspective transformation parameters using either knowledge about the internal geometry of the camera or by *camera calibration* techniques (23) in which case the problem is reduced to a rigid registration problem.

2.4 Elastic or fluid transformation

In a number of registration problems it is necessary to account for morphological differences across individuals or changes over time (i.e. pre- and postoperative images). In these cases, transformations are used which allow elastic or fluid deformations. Collins *et al.* (24) use a combination of local rigid transformations for the non-linear alignment of images. Another example comprises so-called thin-plate splines which were first used for the point-based elastic registration by Bookstein (25). Thin-plate splines produce a smooth non-linear transformation by interpolating the location of point landmarks using a linear combination of radial basis functions. A number of similar transformation schemes based on different radial basis functions have been proposed (26, 27). Other approaches use physics-based elastic transformation schemes and are closely related to the concept of deformable models which use *a priori* information, such as physical constraints. A recent review of the literature is provided by McInerney and Terzopoulos (28). One of the earliest examples for elastic registration has been proposed by Bajcsy and Kovačič (29) and simulates a rubber-like deformation without tearing or folding. Other deformations are based on fluid deformations (9, 30). A closely related topic is the regularization of elastic or fluid transformations. These regularization terms act as forces which prevent unrealistic deformations. In elastic deformations physical quantities like stress or strain are often used. In fluid deformations, a regularization can be achieved by modelling the object as fluid with a certain viscosity.

3. Registration basis

Once we have defined the transformation which relates the two images, we have to define a registration basis to relate both images. We can classify the registration process into three categories depending on the features used as registration basis: points, contours/surfaces, and voxels.

3.1 Point-based registration

One of the most intuitive registration criteria is the proximity of corresponding point landmarks which can be identified in the two images. Given two point sets **p** and **p**′, we can define a similarity measure based on the squared distance of the corresponding points:

$$S = - \sum_i (\mathbf{p}'_i - \mathbf{T}(\mathbf{p}_i))^2 \qquad [5]$$

In the case of rigid-body transformations, three or more non-collinear landmarks are sufficient to establish the transformation between two 3D image volumes. The algorithm for direct computation of the optimal transformation is well known and straightforward (31). It involves the alignment of centroids of the two sets of points followed by a rotation to minimize the sum of the squared displacements between source and destination points. This is achieved by simple matrix manipulation using the method of singular value decomposition (SVD) (32).

The landmarks may be extrinsic point-like features such as pins or markers fixed to the patient and visible on each scan. These may be attached to the skin or screwed into the bone. The latter can provide very accurate registration but are more invasive and uncomfortable for the patient. Skin markers on the other hand can easily move by several millimetres due to the mobility of the skin and are difficult to attach firmly. Care must be taken to ensure that the coordinate of each marker is computed as accurately as possible and that the coordinate computed in each modality corresponds to the same point in physical space. Sub-voxel precision is possible, for example using the intersection of two lines (33), the apex of a 'V' (34) and the centre of gravity of spherical or cylindrical markers with a volume much larger than the voxel sizes (35). The task of identifying the markers can be automated using computer vision techniques (36).

Alternatively, corresponding intrinsic point-like features such as anatomical landmarks may be identified interactively on each image. These may correspond to truly point-like structures (e.g. the apical turn of the cochlea), structures in which points can be unambiguously defined (e.g. the junction of the vertebral arteries or the centre of the orbit of the eyes). Other possibilities include the identification of geometrical features of anatomical structures such as a maximum of 2D curvature (e.g. the cartoid syphon) or a maximum of 3D curvature (e.g. the occipital pole of the brain). The use of corresponding anatomical landmarks that have been identified interactively by a skilled user has been widely reported for the registration of clinical images (4, 37). Registration errors are reduced by increasing the number of fiducial markers. If the error in landmark identification is randomly distributed about the true landmark position, the error in coordinates computed from the resulting transformation reduces as the square root of the number of points identified, for a given distribution of points. Finding landmarks automatically and reliably is difficult and remains a research issue. Therefore the majority of point-based registration systems rely on the manual identification of landmarks in an interactive fashion. Furthermore, care has to be taken to select only those landmarks which are rigidly attached to the structures of interest. In order to avoid registration errors, the plausibility of the registration solution should be verified by calculating error measures such as the root mean square and maximum error between corresponding points. This provides an important feedback of information to the user. The point-based registration procedure is summarized in *Protocol 2*.

Protocol 2. A point-based registration procedure

1. Define and localize two sets of corresponding points by
   (a) identifying extrinsic landmarks such as skin or bone implanted markers and/or
   (b) identifying intrinsic landmarks such as anatomical or geometrical landmarks.

Protocol 2. *Continued*

2. Compute the rigid transformation parameters between both set of points by minimizing the squared distance between corresponding points:

    (a) calculate the translation parameters by aligning the centroids of both points sets;

    (b) calculate the rotation parameters using the SVD algorithm proposed by Arun *et al.* (31).

3. Verify the plausibility of the registration transformation by calculating

    (a) the maximum error for corresponding points and/or

    (b) the root mean square (RMS) error for corresponding points.

Figure 1 shows an example CT volume registered with a gadolinium-enhanced MR image by picking point landmarks. The spatial relationships between bone and tumour, an acoustic neuroma, are clearly seen. Also shown is a volume rendered image of the combined datasets registered with an additional MR angiogram acquired to demonstrate the relationship between the vasculature and tumour (38). The vasculature is shown in red and the tumour in green from the different MR acquisitions, and bone is shown in grey from the CT. The slices are viewed according to radiographic convention from below, while the rendering is generated as if viewing the skull and its contents from above. *Figure 2* shows an example of aligned and overlaid MR images (grey) and PET 18-fdg images (green) of the brain (*Figure 2a*) and aligned and overlaid CT images (grey) and PET 18-fdg images (green) of the pelvis (*Figure 2b*), showing separation of residual bladder activity from the recurrent carcinoma.

3.2 Contour- and surface-based registration

Another feature which can be used for the registration of images consists of the boundaries of corresponding anatomical structures. This requires the segmentation of a contour or surface representing the object of interest in both images. This can be achieved using an interactive or automated segmentation procedure. The contours or surfaces of both objects are generally represented as point sets which can be registered by minimizing the distance between corresponding points of both sets (39, 40). Since the correspondence between both point sets is not known a priori, Besl and McKay (41) proposed a generic registration algorithm, called the *Iterative Closest Point* (ICP) algorithm, which assumes that there is a correspondence between each point in the first set and its closest point in the second set. The point sets are then registered using the

6: 3D analysis: registration of biomedical images

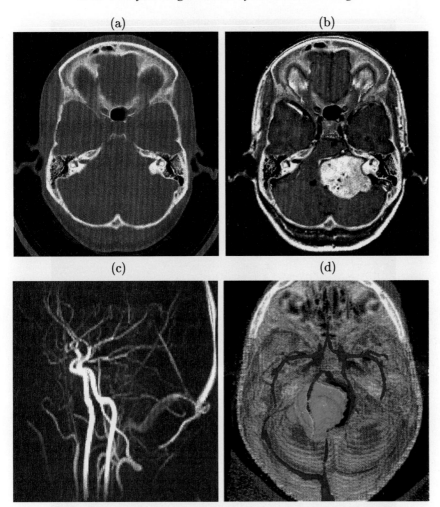

Figure 1. (a) A transaxial slice of an X-ray CT volume. (B) A transaxial slice of this volume registered and overlaid over the corresponding transaxial slice of the gadolinium-enhanced MR volume. (c) A lateral projection of the vasculature from an MR angiogram. (d) A rendering derived from all three image volumes. The bone in grey is derived from the CT, the tumour in green from the gadolinium-enhanced MR, and the vasculature in red from the MR angiogram.

point-based technique discussed in the previous section. The process is repeated until convergence is achieved. The ICP registration procedure is described in *Protocol 3*. In practice, the computational cost can be significantly reduced by using distance transforms such as the Chamfer distance (42) to pre-calculate the distance to the closest point.

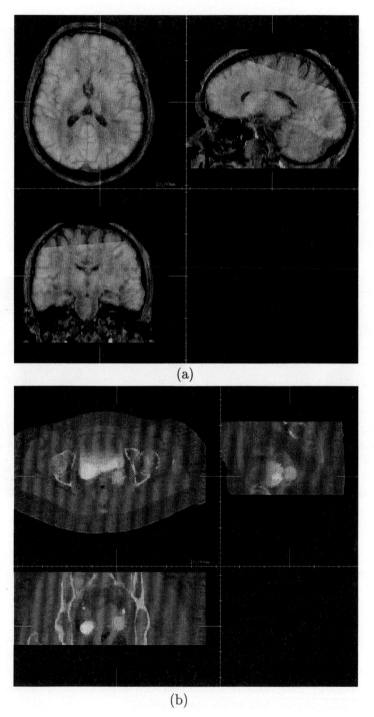

(a)

(b)

Figure 2. (a) Three orthogonal slices through registered and combined PET 18-fdg and MR volumes of the head, with the PET image overlaid in green on the MR image displayed on a grey scale. (b) Superimposition of a transaxial slice from registered CT (grey) and PET 18-fdg (green) image volumes of the pelvis, showing clear separation of residual bladder activity from recurrent cervical carcinoma.

Protocol 3. The *Iterative Closest Point* (ICP) registration procedure

1. Define and localize two sets of points by identifying corresponding contours or surfaces.
2. For each point in the first point set find the closest point in the second point set and label both points as corresponding points.
3. Compute the rigid transformation parameters between both set of points assuming the point corresondences calculated in the previous step. The rigid transformation parameters can be calculated as described in *Protocol 2*.
4. Apply the rigid transformation to the second point set.
5. Compute the distance between corresponding points. If the distance is larger than a predefined threshold go to step 2 and repeat the procedure, otherwise stop.

In general, contour- and surface-based registration techniques use more of the available information than the point-based approach and are therefore often more robust and accurate. However, the accuracy of these techniques is highly dependent on the accurate identification of corresponding surfaces, yet different imaging modalities can provide very different image contrast between different structures. The process of delineation is arduous to do accurately. Computer assisted segmentation will almost always require some manual editing or adjustment. Contours and surfaces may also exhibit natural symmetries to certain rotations leading to poorly constrained transformations. The use of geometric features is not restricted to contours or surfaces. Other features which can be represented as point sets such as lines and tubes have also been used (43, 44). In addition, it is possible to use a weighted combination of different features. For example, Maurer *et al.* (45) have shown that the registration of CT and MR images using points and surfaces is significantly more accurate than using either points or surfaces. We have shown how, in principle, adjacent surfaces may be used for registration incorporating knowledge of the spatial relationships of different surfaces (46).

3.3 Voxel-based registration

Both registration techniques described previously rely on the identification of corresponding points, contours, or surfaces. In most cases this requires a segmentation step and can limit the accuracy of the registration. The aim to avoid the necessity of the segmentation step has led to the development of voxel-based similarity measures. These voxel-based similarity measures use the information provided by the image intensities directly and are motivated by the observation that while images from different modalities exhibit complementary information, there is usually a high degree of shared information between images of the same structures. The application of voxel-based similarity measures for mono- and multimodal registration tasks has shown promising results in terms of robustness and accuracy.

3.3.1 Statistical measures

The simplest statistical measure of image similarity is based on the squared sum of intensity differences (SSD) between $\Im_A$ and $\Im_B$,

$$\mathfrak{S}_{\text{SSD}} = -\frac{1}{n}\sum(\Im_A(\mathbf{p'}) - \Im_B(\mathbf{T}(\mathbf{p})))^2 \qquad [6]$$

where n is the number of voxels in the region of overlap. This measure is based on the assumption that both imaging modalities have the same characteristics. If the images are correctly aligned, the difference between them should be zero except for the noise produced by the two modalities. If this noise is Gaussian distributed, it can be shown that the SSD is the optimal similarity measure (47). Since this similarity measure assumes that the imaging modalities are identical, their application is restricted to monomodal applications such as serial registration (13, 14).

In a number of cases, the assumption of identical imaging modalities is too restrictive. A more general assumption is that of a linear relationship between the two images. In these cases, the similarity between both images can be expressed by the normalized cross-correlation (NCC)

$$\mathfrak{S}_{\text{NCC}} = \frac{\sum \Im_A(\mathbf{p'}) - \mu_A)(\Im_B(\mathbf{T}(\mathbf{p})) - \mu_B)}{[(\sum \Im_A(\mathbf{p'}) - \mu_A)^2(\sum \Im_B(\mathbf{T}(\mathbf{p'})) - \mu_B)^2]^{\frac{1}{2}}} \qquad [7]$$

where μ_A, μ_B correspond to average voxel intensities in both images. Nevertheless, the application of this similarity measure is largely restricted to monomodal registration tasks. To apply correlation in multimodal registration tasks such as MR to CT, van den Elsen et al. (48) proposed a method based on the remapping of the CT image intensities using a triangular look-up table so that both air and bone appear dark while soft tissue appears bright, as in the corresponding MR image. Successful registration of images of the head and spine were reported. Another approach is based on the correlation of geometric features derived from the image intensities rather than the image intensities themselves. Possible geometric features include edges (49) or ridges (50, 51) of the image intensities.

One of the earliest successful applications of voxel similarity measures was the variance of intensity ratios (VIRs) proposed by Woods et al. (52) for the registration of MR and PET images of the brain. Here, it is assumed that the variance of corresponding PET intensities for all voxels within defined ranges of MR intensity is minimized at registration. The approach works very well if the brain is segmented manually by removing non-brain regions (i.e. skin, skull, and surrounding air) from the MR image. The ratio of variance is calculated as follows: for each voxel value a in image $\Im_A$ the normalized standard deviation of the corresponding voxels in image $\Im_B$ is calculated as the ratio of the mean $\mu_B(a)$ and standard deviation $\sigma_B(a)$. The ratio of variance is then calculated as the weighted mean of the normalized standard deviations

$$\mathfrak{S}_{VIR} = \sum_{a \in \Im_A} \frac{n_A(a)}{n} \frac{\sigma_B(a)}{\mu_B(a)} \qquad [8]$$

where $n_A(a)$ is the number of voxels with intensity a in image $\Im_A$, and n is the total number of voxels.

3.3.2 Information-theoretical measures

There has been significant interest in measures of alignment based on the information content or entropy of the registered image. An important step to understanding these methods is the feature space of the image intensities which can also be interpreted as the joint probability distribution (53). An example of the feature space of an MR and CT image of the head at alignment and at two misalignments is shown in *Figure 3*. Similarly, the feature space of an MR and PET image of the same patient is shown in *Figure 4*. The feature space is formed by accumulating a two-dimensional histogram of the co-occurrences of intensities in the two images for each trial alignment. It can be seen that the feature space disperses as misalignment increases, and that each image pair has a distinctive feature space signature at alignment.

Minimizing the entropy of the joint probability distribution minimizes the dispersion of the feature space and has been proposed by Studholme *et al.*

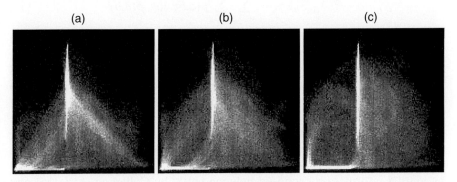

Figure 3. The feature spaces of an MR and CT image (a) at registration, (b) misregistered by 2 mm, and (c) misregistered by 5 mm. The vertical axes correspond to the MR image intensity and the horizontal axes to CT image intensity.

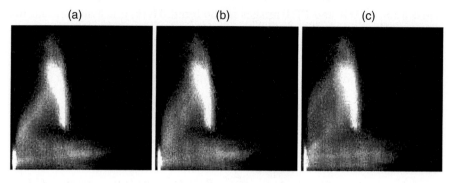

Figure 4. The feature spaces of an MR and PET image (a) at registration, (b) misregistered by 2 mm, and (c) misregistered by 5 mm. The MR intensities are plotted on the vertical axes, and the PET intensities are plotted on the horizontal axes.

(54) as a measure of misalignment. As an illustrative example, consider two images of the same individual, each containing two eyes. Misaligned, the combined images will contain four eyes while at alignment there will only be two eyes. There is, therefore, less information in the combined images at registration.

The Shannon–Wiener entropies, $H(\Im_A)$ and $H(\Im_B)$, of images $\Im_A$ and $\Im_B$ may be defined by

$$H(\Im_A) = - \sum_{a \in \Im_A} p(a) \log p(a) \qquad [9]$$

and

$$H(\Im_B) = - \sum_{a \in \Im_B} p(b) \log p(b) \qquad [10]$$

where $p(a)$ is the probability that a voxel in image $\Im_A$ has intensity a and $p(b)$ is the probability that a voxel in image $\Im_B$ has intensity b. The joint entropy $H(\Im_A, \Im_B)$ of the overlapping region of images $\Im_A$ and $\Im_B$ may be defined by

$$H(\Im_A, \Im_B) = - \sum_{a \in \Im_A} \sum_{a \in \Im_B} p(a,b) \log p(a,b) \qquad [11]$$

where $p(a,b)$ is the joint probability that a voxel in the overlapping region of image $\Im_A$ and $\Im_B$ has values a and b, respectively.

While the joint entropy can provide a useful measure of alignment it has proved not to be very robust, as other misalignments can result in a much lower joint entropy. For example, alignment of just the air surrounding the patient will produce a global minimum of entropy. A more appropriate measure would be the difference in information between the overlapping volume of the combined image with respect to the information in the overlapping volumes of the two original images. Such a measure is provided by mutual information proposed independently by Collignon et al. (55) and Viola and Wells (56). Mutual information (MI) is given by

$$\mathfrak{S}_{MI}(\Im_A; \Im_B) = H(\Im_A) + H(\Im_B) - H(\Im_A, \Im_B) \qquad [12]$$

and it should be maximal at alignment. Mutual information is a measure of how well one image 'explains' the other but makes no assumption of the functional form or relationship between image intensities in the two images. It has been shown by Studholme *et al.* (57) that mutual information itself is not independent of the overlap between two images. To avoid any dependency on the amount of image overlap, Studholme *et al.* suggested the use of normalized mutual information (NMI) as a measure of image alignment:

$$\mathfrak{S}_{\text{NMI}}(\mathfrak{I}_A;\mathfrak{I}_B) = \frac{H(\mathfrak{I}_A) + H(\mathfrak{I}_B)}{H(\mathfrak{I}_A,\mathfrak{I}_B)} \quad [13]$$

Similar forms of normalized mutual information have been proposed by Maes *et al.* (58).

Entropy-based voxel similarity measures are based on the notion of the marginal and joint probability distributions of the two images. These probability distributions can be estimated in two different ways: the first method uses histograms whose bins count the frequency of occurrence (or co-occurrence) of intensities (shown in *Figures 3* and *4*). Dividing these frequencies by the total number of voxels yields an estimate of the probability of that intensity. The second method is based on generating estimates of the probability distribution using Parzen windows (59) which is a non-parametric technique to estimate probability densities. The Parzen-based approach has the advantage of providing a differentiable estimate of mutual information which is not the case of the histogram-based estimate of mutual information. It should be pointed out that other similarity measures like the sums of squared differences or correlation can be expressed in terms of probability distributions. A comparative overview of the different voxel similarity measures is shown in *Table 1*.

4. Optimization

The various similarity measures discussed in the previous section provide measures of alignment. However, in general it is not possible to calculate the

Table 1. Overview of the properties of different voxel-based similarity measures[a]

Similarity measure	Monomodal	Multimodal	Differentiable	Segmentation
SSD	+	–	+	–
NCC	+	–	+	–
VIR	+	+	+	◊
MI	+	+	◊	–

[a]Properties satisfied or not are indicated by '+' or '–'. Properties which are only partially fulfilled are denoted by '◊' (see text for details). The four categories are (a) applicability to monomodal registration, (b) applicability to multimodal registration, (c) differentiability of the similarity measure, and (d) the need for pre-segmentation of the image.

best estimate of alignment directly. Instead, an iterative optimization process is normally used to compute the best registration estimate. There are a number of standard optimization techniques including downhill descent, downhill simplex, or Powell's method. These techniques do not require any derivative information. In contrast to that, techniques such as steepest or conjugate gradient descent as well Levenberg–Marquardt optimization do require derivative information. An excellent overview of these optimization techniques together with example implementations can be found in Press *et al.* (32). Which of these techniques is the best optimization strategy for specific similarity measures is still a research topic. All of the above techniques can be used within multiresolution schemes which can significantly reduce the computational complexity of the optimization problem and increase the robustness towards local optima. In such a multiresolution scheme the optimization is applied to the images at different resolutions starting at the coarsest resolution. If no further improvement is possible, the found solution is propagated to the next finest level of resolution and the optimization is repeated. The algorithm stops at the finest resolution if no further improvement is possible. It has been shown that such a multiresolution approach in conjunction with simple downhill optimization techniques can yield accurate and robust registration estimates (60).

5. Applications

5.1 2D registration

The registration of 2D images is normally computationally less complex than many 3D registration tasks. This is because 2D transformations have fewer degrees of freedom than their 3D counterparts. For example, the rigid registration of 2D images involves finding a rotation together with two translation parameters. Furthermore, the amount of data involved in the registration process is limited. Nevertheless, the registration of 2D images has a number of important applications. One such application is the registration of X-ray radiographs of the hand for the diagnosis of scaphoid injury (61). Other important applications include the registration of histological images to an atlas (62) or the registration of portal X-ray images (63).

5.2 3D–3D rigid registration

Rigid registration algorithms based on mutual information and multiresolution optimization schemes have been successfully used for the alignment of clinically acquired MR to PET and MR to CT image pairs of the head. Detailed results have been reported by Studholme *et al.* (64, 65), and Studholme (65) showed that the algorithm was robust for initial misalignments of up to 30 mm and 30°, provided sufficient axial sections of image data were available (30 mm or more of axially overlapping images at registration). Registration takes between 5 and 15 min on a medium range Unix workstation, depending on the image size and resolution. *Figure 5* shows example MR and CT images registered fully automatically by this algorithm. The algorithm is now in clinical use in a number of different sites. Our experience has also shown that many clinical imaging systems, especially MR, produce errors of the voxel dimensions of up to a few percent. This can lead to problems especially in

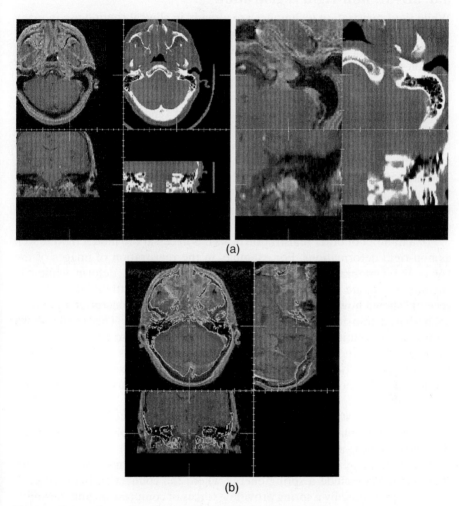

Figure 5. Transaxial and coronal slices through CT and MR volumes of the head, registered fully automatically by multiresolution maximization of mutual information. (a) The panels on the left show side-by-side display of the aligned images and those on the right are zoomed versions of the same images. (b) Alternative display with a contour corresponding to the surface of bone, obtained by intensity thresholding the CT volume, superimposed on the MR images. Note that despite the limited axial extent of the CT volume, the algorithm provides an accurate registration.

serial registration applications where the goal is to detect subtle changes over time. It has been shown by Hill et al. (21) that the registration error can be reduced significantly by using a registration algorithm which determines the scaling as well as rigid parameters.

5.3 3D–3D non-rigid registration

While rigid registration techniques are already in clinical routine use in many sites, the majority of non-rigid registration techniques are still in the stage of development and evaluation. There are a number of different approaches for the non-rigid registration for applications such as the registration of pre- and post-operative images, as well as for the registration of images which contain deformable structures. One such example is the registration of contrast-enhanced MR images for the detection of breast cancer. We have developed an approach for the non-rigid registration which uses free-form deformations (FFDs) based on B-splines, a well-known computer graphics technique, to model the deformations of the breast by maximizing mutual information (66). The algorithm has been applied to the fully automated registration of 3D breast MR in volunteers and patients. The results have shown that this non-rigid registration algorithm is much better able to recover the motion and deformation of the breast than rigid or affine registration algorithms.

For a number of other registration tasks it is necessary to model rigid as well as non-rigid deformations. For example, in the registration of images of the spine, the vertebrae of the spine are rigid and do not deform while the surrounding tissue can deform in a non-rigid fashion. Little et al. (67) have recently shown how the constraints of rigid bodies can be incorporated into a spline-based transformation using radial basis functions. This results in an interpolating solution that is a summation of a linear term corresponding to the rigid bodies and a basis function which smoothly tends to zero at the surface of the rigid bodies. The resulting transformation is exact at rigid bodies, given the rigid body transformation, and provides smooth interpolation elsewhere.

An alternative strategy is to attempt to model soft tissue deformation in a physically more plausible way. We have demonstrated plausible soft tissue deformation using a simple multiresolution model in which the tissue is represented as an array of discrete elements (68). Energy terms associated with these elements include a spring energy where the connection between each node is represented by a spring providing forces of compression and tension; a stiffness energy associated with bending of the connections of sets of three adjacent nodes; a membrane energy associated with changes in the area of triangular elements; and combinations of all three. This method has been tested on 2D CT and MR slices of the brain acquired before and after surgery for placement of electrode mats on the brain surface prior to excision of areas of focal activity in the treatment of epilepsy. We modelled the scanned slice with a three-component model, one component representing rigid structures (bone), one representing fluid (cerebrospinal fluid and air surrounding the patient) which has zero energy associated with deformation, and the third representing deforming soft tissues. Here the image data provided their own internal standard, in this case electrodes placed in the inter-hemispheric fissure, and the combination of the stiffness and membrane model proved to be most accurate. Results are shown in *Figure 6*.

In general, non-rigid registration techniques can increase the computational complexity of the registration algorithm significantly. The reasons for this are two-fold: Firstly, non-rigid transformations are normally more complex to calculate than rigid transformations. This increases the computational time for each evaluation of the similarity measure. Secondly, non-rigid transformations have usually a much higher number of degrees of freedom. This leads to a significant increase of complexity for the optimization procedure and requires a much higher number of evaluations of the similarity measure. For example the rigid registration (6 degrees of freedom) of breast MR images takes 1–2 min computing time on a Sun Ultra 2 workstation. The computing time for an affine registration (12 degrees of freedom) increases to 2–5 min while the computing time for a non-rigid registration with several thousand degrees of freedom increases to 30–60 min.

5.4 Validation

Any process that entails manipulation of data for clinical purposes must undergo extensive validation. In the case of image registration, this will usually involve a sequence of evaluations on computer-generated models (software phantoms), images of physical phantoms of accurately known construction and dimensions, and images of patients or volunteers. The process must demonstrate both high robustness and high accuracy. Robustness implies a very low failure rate and, if failure does occur, that this is communicated to the user.

Assessment of accuracy requires knowledge of a *gold standard* or *ground truth* registration. This is difficult to achieve with clinical images, but three methods have recently been reported. In one approach one image is simulated from another. Strother *et al.* (70) generated PET images from six MR images of the head using the characteristics of the CTI-Siemens 953B PET scanner. They tested five algorithms including using a stereotactic frame, user-identified anatomical landmarks, surface matching, and Woods algorithm for MR to MR registration, MR to PET registration, and PET to PET registration. Woods algorithm, the only voxel similarity technique tested, performed best. Hemler *et al.* (71) used a cadaver head in which glass tubes had been inserted to provide a gold standard registration of MR and CT images. They compared surface-, stereotactic frame-, and voxel intensity-based correlation, reporting that stereotactic frame-based registration was the most accurate.

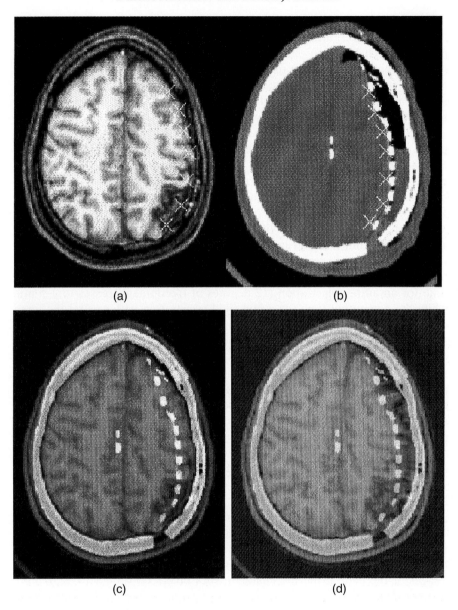

Figure 6. Slice of a pre-operative MR volume of a patient suffering from epilepsy (a). The corresponding CT slice is shown (b) after craniotomy and placement of the electrode mat. Rigid body registration shows (c) significant displacement of the surface of the brain which is corrected, in this case, with a deformation computed from a combined stiffness and membrane model (d) (from ref. 69).

By far the most extensive study to date is that undertaken by Vanderbilt University (72). In this study seven sets of MR and CT images and seven sets of MR and PET images were made available to researchers participating in the study. The patients imaged had marker pins inserted into the bone of their skulls prior to imaging for image-guided neurosurgery. These markers provided a very accurate *gold standard* registration. Prior to distribution the markers were digitally removed from the images. The registrations were undertaken at each site blind to gold standard estimates. A number of surface- and voxel-based methods were compared and the voxel-based methods performed significantly better and required little or no user interaction. The results using mutual information were amongst the most accurate. The median registration errors for the MR and CT images were 1.9 mm or better and for the MR and PET images were 3.2 mm or better using our implementation. In the absence of any independent gold standard, the registration results can be validated by visual assessment. Fitzpatrick *et al.* (73) have shown that experienced observers are sensitive to registration errors of more than 2 mm.

All these validation methods have only been applied to the head, and only in situations where the rigid body transformation is a good approximation. The validation of algorithms which incorporate deformations is difficult and remains a research issue. Deformations to account for variations across populations have potentially a very large number of degrees of freedom while deformations during surgical interventions will vary significantly from patient to patient. We need to demonstrate that the transformations predicted by either of our two approaches are more accurate than those provided by the conventional rigid body assumptions. This could be done using anatomical landmarks, but it is difficult to define landmarks which can be identified accurately. Paradoxically, if these landmarks could be identified then they should be used to aid registration. Alternatives include the use of cadavers or animals, both of which pose significant logistic and ethical problems. It might be possible to simulate imaging processes using the widely available Visible Human Dataset (74). This dataset consists of high-resolution MR and CT images all from the same individual, obtained postmortem.

6. Summary

In this chapter, we have introduced a number of different registration techniques for the alignment of biomedical images. Many of the methodologies described here have applications beyond medicine and biology. We have described the basic components of any registration procedure, namely the registration transformation, the registration basis, and the optimization strategy. It is important to bear in mind that the specific implementations of these components are highly dependent on the particular registration problem.

The registration transformation forms an integral part of any image alignment procedure. Rigid registration techniques are suitable for a large number of intra-subject registration tasks in which the rigid body assumption is adequate and is therefore nowadays widely used in a number of clinical routine applications. In contrast, non-rigid registration techniques have not been used in clinical practice even though they have a number of important applications such as the intra-subject registration of deformable structures as well as the inter-subject registration. The main reasons for this are the computational difficulties surrounding the realistic and plausible modelling of soft-tissue deformations as well as the increasing computational complexity. Moreover, the validation of these algorithms therefore remains a significant problem with further research required to develop appropriate validation strategies.

The other integral part of any image alignment procedure is formed by the registration basis. Registration techniques based on points, contours, or surfaces have the advantage of providing an intuitive approach to image alignment. On the other hand, these techniques often require significant manual interaction. In recent years, voxel-based registration algorithms, in particular those based on information-theoretic similarity measures such as mutual information, have demonstrated their ability for the accurate and robust image alignment without any need for manual interventions. These measures have the advantage that they are easy to implement since they operate directly on the image intensities. Furthermore, they can be applied to mono- and multimodal registration problems. This has led to the fact that voxel-based registration algorithms have become the method of choice for many applications.

Finally, a number of software packages for image registration are available on the Internet. Among these packages is the registration based on the maximization of mutual information which is available at `http://carmen.umds.ac.uk/cisg/software`. Another registration package, called AIR, is based on the minimization of the variance of intensity ratios and is available at `http://bishopw.loni.ucla.edu/AIR3`. Finally, a package for the analysis and registration of SPECT/PET and MR is available at `http://www.fil.ion.ucl.ac.uk/spm`.

Acknowledgements

The work at UMDS described in this paper was funded by EPSRC and Philips Medical Systems. The work was undertaken by Dr Derek Hill, Dr John Little, Dr Colin Studholme, Philip Edwards, Graeme Penney, and Cliff Ruff. The technical support of Colin Renshaw, the radiography staff at Guy's and St. Thomas' Hospitals and King's Healthcare, and the staff at the Clinical PET Centre is acknowledged. The work described here would not have been possible without the support of our clinical and surgical colleagues and in particular Dr Tim Cox, Professor Michael Gleeson, Mr Anthony Strong, Professor Charles Polkey, and Professor Michael Maisey.

References

1. Brown, L. G. (1992). *ACM Computer Surveys*, **24**, 325–375.
2. van den Elsen, P. A., Pol, E.-J. D., and Viergever, M. A. (1993). *IEEE Engineering in Medicine and Biology Magazine*, **12**, 26–39.
3. Maintz, J. B. A. and Viergever, M. A. (1998). *Medical Image Analysis*, **2**(1), 1–36.
4. Hill, D. L. G., Hawkes, D. J., Gleeson, M. J., Cox, T. C. S., Strong, A. J., Wong, W.-L., Ruff, C. F., Kitchen, N. D., Thomas, D. G. T., Crossman, J. E., Studholme, C., Gandhe, A. J., Green, S. E. M., and Robinson, G. P. (1994). *Radiology*, **191**, 447–454.
5. Levin, D. N., Hu, X., Tan, K. K., Galhorta, S., Pelizarri, C. A., Chen, G. T. Y., Beck, R. N., Chen, C. T., Cooper, M. D., Mullan, J. F., Hekmatpanah, J., and Spire, J. P. (1989). *Radiology*, **175**(3), 783–789.
6. Wong, W. L., Hussain, K., Chevretton, E., Hawkes, D. J., Baddeley, H., Maisey, M., and McGurk, M. (1996). *American Journal of Surgery*, **172**(6), 628–632.
7. Collins, D. L., Neelin, P., Peters, T. M., and Evans, A. C. (1994). *Journal of Computer Assisted Tomography*, **18**(2), 192–205.
8. Friston, K. J., Frith, C. D., Liddle, P. F., and Frackowiak, R. S. J. (1997). *Journal of Cerebral Blood Flow and Metabolism*, **11**, 690–699.
9. Christensen, G. E., Miller, M. I., Mars, J. L., and Vannier, M. W. (1995). In *Computer assisted radiology*, pp. 146–151. Springer, Berlin, Germany.
10. Talairach, J. and Tournoux, P. (1988). *Co-planar stereotactic atlas of the human brain: 3-dimensional proportional system: an approach to cerebral imaging.* Georg Thieme, Stuttgart, Germany.
11. Schormann, T., Dabringhaus, A., and Zilles, K. (1995). *IEEE Transactions on Medical Imaging*, **14**(1), 25–35.
12. Hill, D. L. G., Hawkes, D. J., Studholme, C., Summers, P. E., and Taylor, M. G. (1994). In *Proc. of the International Society for Magnetic Resonance in Medicine*, IRMRM, San Francisco, CA. Vol. 2, p. 830.
13. Hajnal, J. V., Saeed, N., Oatridge, A., Williams, E. J., Young, I. R., and Bydder, G. M. (1995). *Journal of Computer Assisted Tomography*, **19**(5), 677–691.
14. Hajnal, J. V., Saeed, N., Soar, E. J., Oatridge, A., Young, I. R., and Bydder, G. M. (1995). *Journal of Computer Assisted Tomography*, **19**(2), 289–296.
15. Freeborough, P. A., Woods, R. P., and Fox, N. C. (1996). *Journal of Computer Assisted Tomography*, **20**(6), 1012–1022.
16. Lemieux, L., Wieshmann, U. C., Moran, N. F., Fish, D. R., and Shorvon, S. D. (1998). *Medical Image Analysis*, **2**(3), 227–242.
17. Maciunas, R. J. (1993). *Interactive image-guided neurosurgery.* American Association of Neurological Surgeons.
18. R. H. Taylor, Lavallée S., Burdea G., and Mösges, R. (ed.) (1995). *Computer integrated surgery.* MIT Press, Cambridge, MA.
19. Lavallée, S. (1995). In *Computer-integrated surgery* (ed. R. H. Taylor, S. Lavallée, G. C. Burdea, and R. Mösges). pp. 77–97. MIT Press, Cambridge, MA.

20. Foley, J., van Dam, A., Feiner, S., and Hughes, J. (1990). *Computer graphics*, 2nd edition. Addison Wesley, Reading, MA.
21. Hill, D. L. G., Maurer, C. R., Jr, Studholme, C., Fitzpatrick, J. M., and Hawkes, D. J. (1998). *Journal of Computer Assisted Tomography*, **22**(2), 317–323.
22. Szelski, R. and Lavallée, S. (1996). In *Int. Journal of Computer Vision*, **18**(2), 171–186.
23. Tsai, R. Y. (1987). *IEEE Journal of Robotics and Automation*, **3**(4), 323–344.
24. Collins, D. L., Peters, T. M., and Evans, A. C. (1994). In *Proc. 3rd International Conference Visualization in Biomedical Computing (VBC'94)*, pp. 180–190. SPIE, Rochester, MN.
25. Bookstein, F. L. (1989). *IEEE Transactions on Pattern Analysis and Machine Intelligence*, **11**(6), 567–585.
26. Arad, N., Dyn, N., Reisfeld, D., and Yeshurun, Y. (1994). *Computer Vision, Graphics, and Image Processing: Graphical Models and Image Processing*, **56**(2), 161–172.
27. Davis, M. H., Khotanzad, A., Flamig, D. P., and Harms, S. E. (1997). *IEEE Transactions on Medical Imaging*, **16**(3), 317–328.
28. McInerney, T. and Terzopoulos, D. (1996). *Medical Image Analysis*, **1**(2), 91–108.
29. Bajcsy, R. and Kovačič, S. (1989). *Computer Vision, Graphics and Image Processing*, **46**, 1–21.
30. Bro-Nielsen, M. (1996). In *Proc. 4th International Conference Visualization in Biomedical Computing (VBC'96)*, Springer-Verlag, Hamburg, pp. 267–276.
31. Arun, K. S., Huang, T. S., and Blostein, S. D. (1987). *IEEE Transactions on Pattern Analysis and Machine Intelligence*, **9**, 698–700.
32. Press, W. H., Flannery, B. P., Teukolsky, S. A., and Vetterling, W. T. (1989). *Numerical recipes in C*, 2nd edition. Cambridge University Press.
33. Colchester, A. C. F., Zhao, J., Holton-Tainter, K. S., Henri, C. J., Maitland, N., and Roberts, P. T. E. (1996). *Medical Image Analysis*, **1**, 73–90.
34. van den Elsen, P. A. and Viergever, M. A. (1991). In *Information Processing in Medical Imaging: Proc. 12th International Conference (IPMI'91)*, Springer-Verlag, Wye, UK, pp. 142–153.
35. Maurer, C. R., Jr, Fitzpatrick, J. M., Wang, M. Y., Galloway, R. L., Jr, Maciunas, R. J., and Allen, G. S. (1997). *IEEE Transactions on Medical Imaging*, **16**, 447–462.
36. Wang, M. Y., Maurer, C. R., Jr, Fitzpatrick, J. M., and Maciunas, R. J. (1996). *IEEE Transactions on Biomedical Engineering*, **43**, 627–637.
37. Hill, D. L. G., Hawkes, D. J., Crossman, J. E. *et al.* (1991). *British Journal of Radiology*, **64**, 1030–1035.
38. Ruff, C. F., Hill, D. L. G., Robinson, G. P., and Hawkes, D. J. (1993). In *Proc. Computer Assisted Radiology (CAR'95)*, Springer-Verlag, Berlin, Germany, pp. 574–582.
39. Pelizzari, C. A., Chen, G. T. Y., Spelbring, D. R., Weichselbaum, R. R., and Chen, C. (1989). *Journal of Computer Assisted Tomography*, **13**, 20–26.
40. Jiang, H., Robb, R. A., and Holton, K. S. (1992). In *Proc. 2nd International Conference on Visualization in Biomedical Computing (VBC'92)*, SPIE, Chapel Hill, NC, pp. 196–213.
41. Besl, P. J. and McKay, N. D. (1992). *IEEE Transactions on Pattern Analysis and Machine Intelligence*, **14**(2), 239–256.

42. Borgefors, G. (1986). *Computer Vision, Graphics, and Image Processing*, **34**, 344–371.
43. Meyer, C. R., Leichtman, G. S., Brunberg, J. A., Wahl, R. L., and Quint, L. E. (1995). *IEEE Transactions on Medical Imaging*, **14**, 1–11.
44. Declerck, J., Subsol, G., Thirion, J.-P., and Ayache, N. (1995). In *Proc. International Conference on Computer Vision, Virtual Reality and Robotics in Medicine (CVRMed'95)*, Springer-Verlag, Nice, France, pp. 153–162.
45. Maurer C. R., Jr, Aboutanos, G. B., Dawant, B. M., Maciunas, R. J., and Fitzpatrick, J. M. (1996). *IEEE Transactions on Medical Imaging*, **15**(6), 836–849.
46. Hill, D. L. G. and Hawkes, D. J. (1994). *Image and Vision Computing*, **12**, 173–178.
47. Viola, P. (1995). *Alignment by maximization of mutual information*. PhD thesis, Massachusetts Institute of Technology. A.I. Technical Report No. 1548.
48. van den Elsen, P. A., Pol, E.-J. D., Sumanaweera, T. S., Hemler, P. F., Napel, S., and Adler, J. R. (1994). In *Proc. 3rd International Conference on Visualization in Biomedical Computing (VBC'94)*, SPIE, Rochester, MN, pp. 227–237.
49. Maintz, J. B. A., van den Elsen, P. A., and Viergever, M. A. (1996). *Medical Image Analysis*, **1**(2), 151–161.
50. van den Elsen, P. A., Maintz, J. B. A., Pol, E.-J. D., and Viergever, M. A. (1995). *IEEE Transactions on Medical Imaging*, **14**(2), 384–396.
51. Maintz, J. B. A., van den Elsen, P. A., and Viergever, M. A. (1996). *IEEE Transactions on Pattern Recognition and Machine Intelligence*, **18**(4), 353–365.
52. Woods, R. P., Mazziotta, J. C., and Cherry, S. R., (1993). *Journal of Computer Assisted Tomography*, **17**, 536–546.
53. Hill, D. L. G., Studholme, C., and Hawkes, D. J. (1994). In *Proc. 3rd International Conference on Visualization in Biomedical Computing (VBC'94)*, Vol. 2359, pp. 205–216. SPIE, Rochester, MN.
54. Studholme, C., Hill, D. L. G., and Hawkes, D. J. (1995). In *Information Processing in Medical Imaging: Proc. 14th International Conference (IPMI'95)*, Kluwer Ilede Berder, France, pp. 287–298.
55. Collignon, A., Maes, F., Delaere, D., Vandermeulen, D., Seutens, P., and Marchal, G. (1995). In *Information Processing in Medical Imaging: Proc. 14th International Conference (IPMI'95)*, Kluwer, Ilede Berder, France, pp. 263–274.
56. Viola, P. and Wells, W. M. (1995). In *Proc. 5th International Conference on Computer Vision (ICCV'95)*, IEEE, Cambridge, MA, pp. 16–23.
57. Studholme, C., Hill, D. L. G., and Hawkes, D. J. (1998). *Pattern Recognition*, **32**(1), 71–86.
58. Maes, F., Collignon, A., Vandermeulen, D., Marechal, G., and Suetens, R. (1997). *IEEE Transactions on Medical Imaging*, **16**(2), 187–198.
59. Duda, R. O. and Hart, P. E. (1973). *Pattern classification and scene analysis*. Wiley, New York.
60. Studholme, C., Hill, D. L. G., and Hawkes, D. J. (1997). *Medical Physics*, **24**(1), 25–35.
61. Hawkes, D. J., Robinson, L., Crossman, J. E., Sayman, H. B., Mistry, R., and Maisey, M. N. (1991). *European Journal of Nuclear Medicine*, **18**, 752–756.
62. Cohen, F. S., Yang Z., Huang Z., and Nissanov, J. (1998). *IEEE Transactions on Biomedical Engineering*, **45**(5), 642–649.
63. Dong, L. and Boyer, A. L. (1996). *Physics in Medicine and Biology*, **41**(4), 697–724.

64. Studholme, C., Hill, D. L. G., and Hawkes, D. J. (1996). *Medical Image Analysis*, **1**(2), 163–175.
65. Studholme, C. (1997). *Measures of 3D medical image alignment*. PhD thesis, United Medical and Dental Schools of Guy's and St Thomas's Hospitals, UK.
66. Rueckert, D., Hayes, C., Studholme, C., Summers, P., Leach, M., and Hawkes, D. J. (1998). In *Lecture notes in computer science*, pp. 1144–1152. Springer, Cambridge, MA.
67. Little, J. A., Hill, D. L. G., and Hawkes, D. J. (1997). *Computer Vision and Image Understanding*, **66**(2), 223–232.
68. Edwards P. J., Hill, D. L. G., Little, J. A., and Hawkes, D. J. (1997). In *Information Processing in Medical Imaging: Proc. 15th International Conference (IPMI'97)*, Springer-Verlag, Pultney, Vermont, pp. 218–231.
69. Edwards P. J., Hill, D. L. G., Little, J. A., and Hawkes, D. J. (1998). *Medical Image Analysis*, **2**(4), 355–367.
70. Strother, S. C., Anderson, J. R., Xu, X.-L., Liow, J.-S., Bonar, D. C., and Rottenberg, D. A. (1994). *Journal of Computer Assisted Tomography*, **18**, 954–962.
71. Hemler, P. F., van den Elsen, P. A., Sumanaweera, T., Napel, S., Drace, J., and Adler, J. R. (1995). In *Information Processing in Medical Imaging: Proc. 14th International Conference (IPMI'95)*, Kluwer, Ilede Berder, France, pp. 251–262.
72. West, J. B., Fitzpatrick, J. M., Wang, M. Y., Dawant, B. M., Maurer, C. R., Jr, Kessler, R. M., Maciunas, R. J., Barillot, C., Lemoine, D., Collignon, A., Maes, F., Suetens, P., Vandermeulen, D., van den Elsen, P. A., Napel, S., Sumanaweera, T. S., Harkness, B., Hemler, P. F., Hill, D. L. G., Hawkes, D. J., Studholme, C., Maintz, J. B. A., Viergever, M. A., Malandain, G., Pennec, X., Noz, M. E., Maguire, G. Q., Jr, Pollack, M., Pelizzari, C. A., Robb, R. A., Hanson, D., and Woods, R. P. (1997). *Journal of Computer Assisted Tomography*, **21**, 554–566.
73. Fitzpatrick, J. M., Hill, D. L. G., Shyr, Y., West, J., Studholme, C., and Maurer, C. R., Jr (1998). *IEEE Transactions on Medical Imaging*, **17**(4), 571–585.
74. Spitzer, V. M., Whitlock, D., Scherzinger, A. L., and Ackerman, M. J. (1995). *Radiology*, **197**, 533.

7

Model-based methods in analysis of biomedical images

TIM COOTES

1. Introduction

Biomedical images usually contain complex objects, which will vary in appearance significantly from one image to another. Attempting to measure or detect the presence of particular structures in such images can be a daunting task. The inherent variability will thwart naive schemes. However, by using models which can cope with the variability it is possible to successfully analyse complex images.

Here we will consider a number of methods where the model represents the expected shape and local greylevel structure of a target object in an image.

Model-based methods make use of a prior model of what is expected in the image, and typically attempt to find the best match of the model to the data in a new image. Having matched the model, one can then make measurements or test whether the target is actually present.

This approach is a 'top-down' strategy, and differs significantly from 'bottom-up' methods. In the latter the image data is examined at a low level, looking for local structures such as edges or regions, which are assembled into groups in an attempt to identify objects of interest. Without a global model of what to expect, this approach is difficult and prone to failure.

A wide variety of model-based approaches have been explored (see the review below). This chapter will concentrate on a statistical approach, in which a model is built from analysing the appearance of a set of labelled examples. Where structures vary in shape or texture, it is possible to learn what are plausible variations and what are not. A new image can be interpreted by finding the best plausible match of the model to the image data. The advantages of such a method are that

- It is widely applicable. The same algorithm can be applied to many different problems, merely by presenting different training examples.
- Expert knowledge can be captured in the system in the annotation of the training examples.

- The models give a compact representation of allowable variation, but are specific enough not to allow arbitrary variation different from that seen in the training set.
- The system needs to make few prior assumptions about the nature of the objects being modelled, other than what it learns from the training set. (For instance, there are no boundary smoothness parameters to be set.)

The models described below require a user to be able to mark 'landmark' points on each of a set of training images in such a way that each landmark represents a distinguishable point present on every example image. For instance, when building a model of the appearance of an eye in a face image, good landmarks would be the corners of the eye, as these would be easy to identify and mark in each image. This constrains the sorts of applications to which the method can be applied—it requires that the topology of the object cannot change and that the object is not so amorphous that no distinct landmarks can be applied. Unfortunately, this makes the method unsuitable in its current form for objects which exhibit large changes in shape, such as some types of cells or simple organisms. The remainder of this chapter will briefly review other model-based methods, describe how one form of statistical model can be built and tested, and how such models can be used to interpret objects in new images.

2. Background

The simplest model is to use a typical example as a 'golden image'. A correlation method can be used to match (or register) the golden image to a new image. If structures in the golden image have been labelled, this match then gives the approximate position of the structures in the new image. For instance, one can determine the approximate locations of many structures in a magnetic resonance (MR) image of a brain by registering a standard image, where the standard image has been suitably annotated by human experts. However, the variability of both shape and texture of most targets limits the precision of this method.

One approach to representing the variations observed in an image is to 'hand-craft' a model to solve the particular problem currently addressed. For instance Yuille *et al.* (1) build up a model of a human eye using combinations of parameterized circles and arcs. Although this can be effective it is complicated, and a completely new solution is required for every application.

Staib and Duncan (2) represent the shapes of objects in medical images using Fourier descriptors of closed curves. The choice of coefficients affects the curve complexity. Placing limits on each coefficient constrains the shape somewhat but not in a systematic way. It can be shown that such Fourier models can be made directly equivalent to the statistical models described below, but are not as general. For instance, they cannot easily represent open boundaries.

Kass et al. (3) introduced active contour models (or 'snakes') which are energy-minimizing curves. In the original formulation the energy has an internal term which aims to impose smoothness on the curve, and an external term which encourages movement towards image features. They are particularly useful for locating the outline of general amorphous objects, such as some cells (see Chapter 3, Section 3.1, for the application of a snake to microscope images). However, since no model (other than smoothness) is imposed, they are not optimal for locating objects which have a known shape. As the constraints are weak, this can easily converge to incorrect solutions.

Alternative statistical approaches are described by Grenander and Miller (4) and Mardia et al. (5). These are, however, difficult to use in automated image interpretation. Goodall (6) and Bookstein (7) use statistical techniques for morphometric analysis, but do not address the problem of automated interpretation. Kirby and Sirovich (8) describe statistical modelling of grey-level appearance (particularly for face images) but do not address shape variability.

A more comprehensive survey of deformable models used in medical image analysis is given in ref. 9.

3. Application

We demonstrate the method by applying it to two problems, that of locating features in images of the human face and that of locating the cartilage in MR images of a knee.

Images of the face can demonstrate a wide degree of variation in both shape and texture. Appearance variations are caused by differences between individuals, the deformation of an individual face due to changes in expression and speaking, and variations in the lighting. Typically, one would like to locate the features of a face in order to perform further processing (see *Figure 1*). The ultimate aim may vary widely, from determining the identity or expression of the person to deciding in which direction they are looking (10).

When analysing the MR images of the knee (*Figure 2*), we wish to accurately locate the boundary of the cartilage in order to estimate its thickness and volume (21).

In both applications difficulties arise because of the complexity of the target images. Bottom-up approaches are unlikely to solve the problems successfully. Complex images give rise to many primitives, which must be linked together correctly to give a successful interpretation. The number of ways in which they can be connected explodes exponentially with the number of components. Thus a bottom-up approach can become inefficient for complex scenes. By building models from sets of examples we can apply the same algorithms to both applications. In each case the output is a set of model

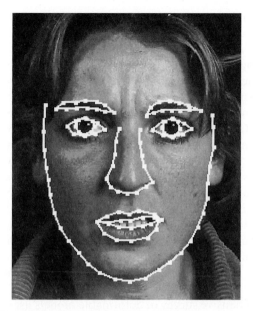

Figure 1. Example face image annotated with landmarks.

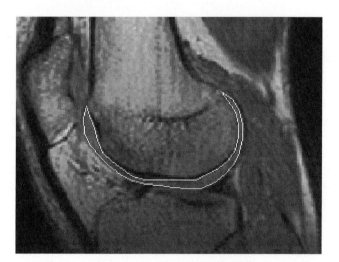

Figure 2. Example MR image of knee with cartilage outlined.

landmark points which best match the image data, together with the model parameters required to generate the points. The points or the parameters can then be used in further processing. For instance, in the face example the parameters can be used to estimate the person's expression or identity. In the knee example, we can estimate the area of the cross-section of the cartilage from the points.

4. Theoretical background

Here we describe the statistical models of shape and appearance used to represent objects in images. A model is trained from a set of images annotated by a human expert. By analysing the variations in shape and appearance over the training set, a model is built which can mimic this variation. To interpret a new image we must find the parameters which best match a model instance to the image. We describe an algorithm which can do this efficiently. Having fit the model to the image, the parameters or the model point positions can be used to classify or make measurements, or as an input to further processing.

4.1 Building models

In order to locate a structure of interest, we must first build a model of it. To build a statistical model of appearance we require a set of annotated images of typical examples. We must first decide upon a suitable set of landmarks which describe the shape of the target and which can be found reliably on every training image.

4.1.1 Suitable landmarks

Good choices for landmarks are points at clear corners of object boundaries, 'T' junctions between boundaries, or easily located biological landmarks. However, there are rarely enough of such points to give more than a sparse description of the shape of the target object. We augment this list with points along boundaries which are arranged to be equally spaced between well-defined landmark points

To represent the shape we must also record the connectivity defining how the landmarks are joined to form the boundaries in the image. This allows us to determine the direction of the boundary at a given point. Suppose the landmarks along a curve are labelled $\{(x_1, y_1), (x_2, y_2), ..., (x_n, y_n)\}$.

For a 2D image we can represent the n landmark points, $\{(x_i, y_i)\}$, for a single example as the $2n$ element vector, **x**, where

$$\mathbf{x} = (x_1, ..., x_n, y_1, ..., y_n)^T. \qquad [1]$$

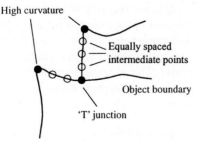

Figure 3. Good landmarks are points of high curvature or junctions. Intermediate points can be used to define the boundary more precisely.

If we have s training examples, we generate s such $\mathbf{x}_j$. Before we can perform statistical analysis on these vectors it is important that the shapes represented are in the same coordinate frame. The shape of an object is normally considered to be independent of the position, orientation, and scale of that object. A square, when rotated, scaled, and translated, remains a square. Appendix A describes how to align a set of training shapes into a common coordinate frame. The approach is to translate, rotate, and scale each shape so that the sum of distances of each shape to the mean ($D = \Sigma|\mathbf{x}_j - \bar{\mathbf{x}}|^2$) is minimized.

4.1.2 Statistical models of shape

Suppose now we have s sets of points $\mathbf{x}_j$ which are aligned into a common coordinate frame. These vectors form a distribution in the $2n$-dimensional space in which they live. If we can model this distribution, we can generate new examples, similar to those in the original training set, and we can examine new shapes to decide whether they are plausible examples.

To simplify the problem, we first wish to reduce the dimensionality of the data from $2n$ to something more manageable. An effective approach is to apply principal component analysis (PCA) to the data. The data form a cloud of points in the $2n$D space, although after aligning the points they lie in a ($2n - 4$)D manifold in this space. PCA computes the main axes of this cloud, allowing one to approximate any of the original points using a model with less than $2n$ parameters (see Appendix C).

If we apply a PCA to the data, we can then approximate any of the training set, $\mathbf{x}$, using

$$\mathbf{x} \approx \bar{\mathbf{x}} + \mathbf{Pb}, \qquad [2]$$

where $\mathbf{P} = (\mathbf{p}_1|\mathbf{p}_2|\cdots|\mathbf{p}_t)$ contains t eigenvectors of the covariance matrix and $\mathbf{b}$ is a t-dimensional vector given by

$$\mathbf{b} = \mathbf{P}^T(\mathbf{x} - \bar{\mathbf{x}}). \qquad [3]$$

(See Appendix B for details.)

The vector $\mathbf{b}$ defines a set of parameters of a deformable model. By varying the elements of $\mathbf{b}$ we can vary the shape, $\mathbf{x}$, using *Equation 2*. The variance of the ith parameter, b_i, across the training set is given by λ_i. By applying limits of $\pm 3 \sqrt{\lambda_i}$ to the parameter b_i we ensure that the shape generated is similar to those in the original training set.

We usually call the model variation corresponding to the ith parameter, b_i, the ith mode of the model. The eigenvectors, $\mathbf{P}$, define a rotated coordinate frame, aligned with the cloud of original shape vectors. The vector $\mathbf{b}$ defines points in this rotated frame.

4.1.3 Examples of shape models

Figure 4 shows example shapes from a training set of 300 labelled faces (see *Figure 1* for an example image showing the landmarks). Each image is

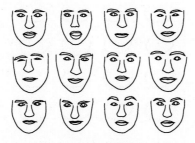

Figure 4. Example shapes from training set of faces.

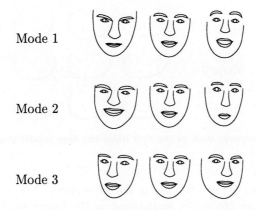

Figure 5. Effect of varying each of the first three face model shape parameters in turn between ±3 SD.

annotated with 133 landmarks. The shape model has 36 parameters, and can explain 98% of the variance in the landmark positions in the training set. *Figure 5* shows the effect of varying the first three shape parameters in turn between ±3 standard deviations from the mean value, leaving all other parameters at zero. These 'modes' explain global variation due to 3D pose changes, which cause movement of all the landmark points relative to one another. Less significant modes cause smaller, more local changes. The modes obtained are often similar to those a human would choose if designing a parameterized model, for instance shaking and nodding the head, or changing expression. However, they are derived directly from the statistics of a training set and will not always separate shape variation in an obvious manner.

Figure 6 shows examples from a training set of 33 knee cartilage boundaries. (See *Figure 2* for an example of the corresponding images.) Each is annotated with 42 landmarks. When we apply the alignment and PCA as described above, we generate a model with 15 modes. *Figure 7* shows the effect of varying the first three shape parameters in turn between ±3 standard deviations from the mean value, leaving all other parameters at zero.

Figure 6. Example shapes from training set of knee cartilages.

Figure 7. Effect of varying each of the first three cartilage model shape parameters in turn between ±3 SD.

Thus in the case of faces the use of the PCA reduced the dimension of the shape vectors from 266 to 36. In the case of the knees, the dimension went from 84 to 15.

4.1.4 Fitting a model to new points

A particular value of the shape vector, **b**, corresponds to a point in the rotated space described by **P**. It therefore corresponds to an example model. This can be turned into an example shape using the transformation from the model coordinate frame to the image coordinate frame. Typically, this will be a euclidean transformation defining the position, (X_t, Y_t), orientation, θ, and scale, s, of the model in the image.

The positions of the model points in the image, **X**, are then given by

$$\mathbf{X} = T_{X_t, Y_t, s, \theta} (\bar{\mathbf{x}} + \mathbf{Pb}) \tag{4}$$

where the function $T_{X_t, Y_t, s, \theta}$ performs a rotation by θ, a scaling by s, and a translation by (X_t, Y_t). For instance, if applied to a single point (x, y),

$$T_{X_t, Y_t, s, \theta} \begin{pmatrix} x \\ y \end{pmatrix} = \begin{pmatrix} X_t \\ Y_t \end{pmatrix} + \begin{pmatrix} s\cos\theta & -s\sin\theta \\ s\sin\theta & s\cos\theta \end{pmatrix} \begin{pmatrix} x \\ y \end{pmatrix}. \tag{5}$$

Suppose now we wish to find the best pose (translation, scale, and rotation) and shape parameters to match a model instance **X** to a new set of image

7: Model-based methods in analysis of biomedical images

points, **Y**. Minimizing the sum of squared distances between corresponding model and image points is equivalent to minimizing the expression

$$|\mathbf{Y} - T_{X_t,Y_t,s,\theta}(\bar{\mathbf{x}} + \mathbf{Pb})|^2. \qquad [6]$$

A simple iterative approach to achieving this is as follows:

Protocol 1. Matching model points to target points

1. Initialize the shape parameters, **b**, to zero (the mean shape).
2. Generate the model point positions using $\mathbf{x} = \bar{\mathbf{x}} + \mathbf{Pb}$.
3. Find the pose parameters (X_t, Y_t, s, θ) which best align the model points **x** to the current found points **Y** (see Appendix A).
4. Project **Y** into the model coordinate frame by inverting the transformation T:

$$\mathbf{y} = T^{-1}_{X_t,Y_t,s,\theta}(\mathbf{Y}). \qquad [7]$$

5. Project **y** into the tangent plane to $\bar{\mathbf{x}}$ by scaling: $\mathbf{y}' = \mathbf{y}/(\mathbf{y} \cdot \bar{\mathbf{x}})$.
6. Update the model parameters to match to **y**′

$$\mathbf{b} = \mathbf{P}^T(\mathbf{y}' = \bar{\mathbf{x}}). \qquad [8]$$

7. If not converged, return to step 2.

Convergence is declared when applying an iteration produces no significant change in the pose or shape parameters. This approach usually converges in a few iterations.

4.1.5 Testing how well the model generalizes

The shape models described use linear combinations of the shapes seen in a training set. In order to be able to generate new versions of the shape to match to image data, the training set must exhibit all the variation expected in the class of shapes being modelled. If it does not, the model will be over-constrained and will not be able to match to some types of new example. For instance, a model trained only on squares will not generalize to rectangles.

One approach to estimating how well the model will perform is to use 'leave-one-out' experiments (see Chapter 4). Given a training set of s examples, build a model from all but one, then fit the model to the example missed out and record the error (for instance using *Equation 6*). Repeat this, missing out each of the s examples in turn. If the error is unacceptably large for any example, more training examples are probably required. However, small errors for all examples only means that there is more than one example for each type of shape variation, not that all types are properly covered (although it is an encouraging sign).

Equation 6 gives the sum of square errors over all points, and may average out large errors on one or two individual points. It is often wise to calculate the error for each point and ensure that the maximum error on any point is sufficiently small.

4.1.6 Choice of number of modes

Section 4.1.2 suggested that during training the number of modes, t, could be chosen so as to explain a given proportion (e.g. 98%) of the variance exhibited in the training set.

An alternative approach is to choose enough modes that the model can approximate any training example to within a given accuracy. For instance, we may wish that the best approximation to an example has every point within one pixel of the corresponding example points.

To achieve this we build models with increasing numbers of modes, testing the ability of each to represent the training set. We choose the first model which passes our desired criteria.

Additional confidence can be obtained by performing this test in a leave-one-out manner. We choose the smallest t for the full model such that models built with t modes from all but any one example can approximate the missing example sufficiently well.

4.2 Image interpretation with models

4.2.1 Overview

To interpret an image using a model, we must find the set of parameters which best match the model to the image. This set of parameters defines the shape and position of the target object in an image, and can be used for further processing, such as to make measurements or to classify the object.

There are several approaches which could be taken to matching a model instance to an image, but all can be thought of as optimizing a cost function. For a set of model parameters, **c**, we can generate an instance of the model projected into the image. We can compare this hypothesis with the target image, to get a fit function $F(\mathbf{c})$. The best set of parameters to interpret the object in the image is then the set which optimizes this measure. For instance, if $F(\mathbf{c})$ is an error measure, which tends to zero for a perfect match, we would like to choose parameters, **c**, which minimize the error measure.

Thus, in theory all we have to do is to choose a suitable fit function, and use a general purpose optimizer to find the minimum. There are many approaches to optimization which can be used, for instance Simplex, Powell's method (12) or genetic algorithms (13). The minimum is defined only by the choice of function, the model, and the image, and is independent of which optimization method is used to find it. However, in practice, care must be taken to choose a function which can be optimized rapidly and robustly, and an optimization method to match.

4.2.2 Choice of fit function

Ideally, we would like to choose a fit function which represents the probability that the model parameters describe the target image object, $P(\mathbf{c}|\mathbf{I})$ (where $\mathbf{I}$ represents the image). We then choose the parameters which maximize this probability.

In the case of the shape models described above, the parameters we can vary are the shape parameters, $\mathbf{b}$, and the pose parameters X_t, Y_t, s, θ.

The form of the fit measure, however, is harder to determine. If we assume that the shape model represents boundaries and strong edges of the object, a useful measure is the distance between a given model point and the nearest strong edge in the image

If the model point positions are given in the vector $\mathbf{X}$, and the nearest edge points to each model point are $\mathbf{X}'$, then an error measure is

$$F(\mathbf{b}, X_t, Y_t, s, \theta) = |\mathbf{X}' - \mathbf{X}|^2. \qquad [9]$$

Alternatively, rather than looking for the best nearby edges, one can search for structure nearby which is most similar to that occurring at the given model point in the training images (see below). It should be noted that this fit measure relies upon the target points, $\mathbf{X}'$, being the correct points. If some are incorrect, due to clutter or failure of the edge/feature detectors, *Equation 9* will not be a true measure of the quality of fit. An alternative approach is to sample the image around the current model points, and determine how well the image samples match models derived from the training set. This approach was taken by Haslam *et al.* (14). A related approach is to model the full appearance of the object, including the internal texture. The quality of fit of such an appearance model can be assessed by measuring the difference between the target image and a synthetic image generated from the model. This is beyond the scope of this chapter, but is described in detail in ref. 15.

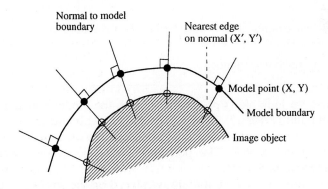

Figure 8. An error measure can be derived from the distance between model points and strongest nearby edges.

4.2.3 Optimizing the model fit

Given no initial knowledge of where the target object lies in an image, finding the parameters which optimize the fit is a difficult general optimization problem. This can be tackled with general global optimization techniques, such as genetic algorithms or simulated annealing (15).

If, however, we have an initial approximation to the correct solution (we know roughly where the target object is in an image, due to prior processing), we can use local optimization techniques such as Powell's method or Simplex. A good overview of practical numeric optimization is given by Press *et al.* (12).

However, we can take advantage of the form of the fit function to locate the optimum rapidly. We derive an algorithm which amounts to a directed search of the parameter space—the active shape model.

4.3 Active shape models

Given a rough starting approximation, an instance of a model can be fit to an image. By choosing a set of shape parameters, **b**, for the model we define the shape of the object in an object-centred coordinate frame. We can create an instance **X** of the model in the image frame by defining the position, orientation, and scale, using *Equation 4*.

An iterative approach to improving the fit of the instance, **X**, to an image proceeds as follows:

Protocol 2. Active shape model algorithm

1. Examine a region of the image around each point X_i to find the best nearby match for the point X'_i.
2. Update the parameters (X_t, Y_t, s, θ, **b**) to best fit the new found points **X**.
3. Apply constraints to the parameters, **b**, to ensure plausible shapes (e.g. limit so $|b_i| < 3\sqrt{\lambda_i}$).
4. Repeat until convergence.

In practise we look along profiles normal to the model boundary through each model point (*Figure 9*). If we expect the model boundary to correspond to an edge, we can simply locate the strongest edge (including orientation if known) along the profile. The position of this gives the new suggested location for the model point.

However, model points are not always placed on the strongest edge in the locality—they may represent a weaker secondary edge or some other image structure. The best approach is to learn from the training set what to look for

7: Model-based methods in analysis of biomedical images

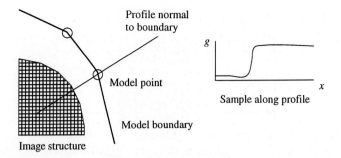

Figure 9. At each model point sample along a profile normal to the boundary.

in the target image. This is done by sampling along the profile normal to the boundary in the training set, and building a statistical model of the greylevel structure.

Suppose for a given point we sample along a profile k pixels either side of the model point in the ith training image. We have $2k + 1$ samples which can be put in a vector $\mathbf{g}_i$. To reduce the effects of global intensity changes we sample the derivative along the profile, rather than the absolute greylevel values. We then normalize the sample by dividing through by the sum of absolute element values,

$$\mathbf{g}_i \rightarrow \frac{1}{\Sigma_j |g_{ij}|} \mathbf{g}_i \qquad [10]$$

We repeat this for each training image, to get a set of normalized samples $\{\mathbf{g}_i\}$ for the given model point. We assume that these are distributed as a multivariate gaussian, and estimate their mean $\bar{\mathbf{g}}$ and covariance $\mathbf{S}_g$. This gives a statistical model for the greylevel profile about the point. This is repeated for every model point, giving one greylevel model for each point.

The quality of fit of a new sample, $\mathbf{g}_s$, to the model is given by

$$f(\mathbf{g}_s) = (\mathbf{g}_s - \bar{\mathbf{g}})^T \mathbf{S}_g^{-1} (\mathbf{g}_s - \bar{\mathbf{g}}) \qquad [11]$$

This is the Mahalanobis distance of the sample from the model mean, and is linearly related to the log of the probability that $\mathbf{g}_s$ is drawn from the distribution. Minimizing $f(\mathbf{g}_s)$ is equivalent to maximizing the probability that $\mathbf{g}_s$ comes from the distribution.

During search we sample a profile m pixels either side of the current point ($m > k$). We then test the quality of fit of the corresponding greylevel model at each of the $2(m - k) + 1$ possible positions along the sample (*Figure 10*) and choose the one which gives the best match (lowest value of $f(\mathbf{g}_s)$).

This is repeated for every model point, giving a suggested new position for each point. We then apply one iteration of the algorithm given in Section 4.1.4 to update the current pose and shape parameters to best match the model to the new points.

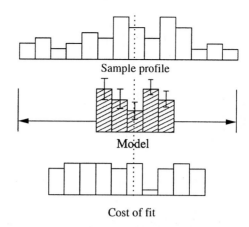

Figure 10. Search along sampled profile to find best fit of greylevel model.

4.3.1 Multiresolution active shape models

To improve the efficiency and robustness of the algorithm, it is implemented in a multiresolution framework. This involves first searching for the object in a coarse image, then refining the location in a series of finer resolution images. This leads to a faster algorithm, and one which is less likely to get stuck on the wrong image structure.

For each training and test image, a gaussian image pyramid is built (17). The base image (level 0) is the original image. The next image (level 1) is formed by smoothing the original, then sub-sampling to obtain an image with half the number of pixels in each dimension. Subsequent levels are formed by further smoothing and sub-sampling (*Figure 11*).

During training we build statistical models of the greylevels along normal profiles through each point, at each level of the gaussian pyramid. We usually use the same number of pixels in each profile model, regardless of level. Since

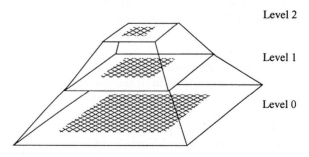

Figure 11. A gaussian image pyramid is formed by repeated smoothing and sub-sampling.

7: Model-based methods in analysis of biomedical images

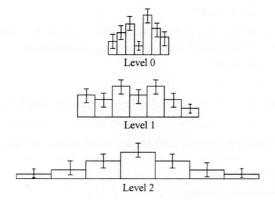

Figure 12. Statistical models of greylevel profiles represent the same number of pixels at each level.

the pixels at level L are 2^L times the size of those of the original image, the models at the coarser levels represent more of the image (*Figure 12*).

Similarly, during search we need only search a few pixels, (n_s), either side of the current point position at each level. At coarse levels this will allow quite large movements, and the model should converge to a good solution. At the finer resolution we need only modify this solution by small amounts.

When searching at a given resolution level, we need a method of determining when to change to a finer resolution, or to stop the search. This is done by recording the number of times that the best found pixel along a search profile is within the central 50% of the profile (i.e. the best point is within $n_s/2$ pixels of the current point). When a sufficient number (e.g. $\geq 90\%$) of the points are so found, the algorithm is declared to have converged at that resolution. The current model is projected into the next image and run to convergence again. When convergence is reached on the finest resolution, the search is stopped.

To summarize, the full multiresolution active shape model (MRASM) search algorithm is as follows:

Protocol 3. Multiresolution search algorithm

1. Set $L = L_{max}$.
2. While $L \geq 0$
    (a) Compute model point positions in image at level L.
    (b) Search at n_s points on profile either side each current point.
    (c) Update pose and shape parameters to fit model to new points.

Protocol 3. *Continued*

    (d) Return to step 2a unless more than p_{close} of the points are found close to the current position, or N_{max} iterations have been applied at this resolution.

    (e) If $L > 0$ then $L \rightarrow (L-1)$.

3. Final result is given by the parameters after convergence at level 0.

The model building process only requires the choice of three parameters:

Model parameters

  n Number of model points
  t Number of modes to use
  k Number of pixels either side of point to represent in grey-model

The number of points is dependent on the complexity of the object, and the accuracy with which one wishes to represent any boundaries. The number of modes should be chosen so that a sufficient amount of object variation can be captured (see Section 4.1.5). The number of pixels to represent in each local grey-model will depend on the width of the boundary structure (however, using between 3 and 7 pixels either side has given good results for many applications).

The search algorithm has four parameters:

Search parameters (suggested default)

  L_{max} Coarsest level of gaussian pyramid to search
  n_s Number of sample points either side of current point (2)
  N_{max} Maximum number of iterations allowed at each level (5)
  p_{close} Desired proportion of points found within $n_s/2$ of current position (0.9)

The levels of the gaussian pyramid to search will depend on the size of the object in the image.

4.3.2 Examples of search

Figure 13 demonstrates using the ASM to locate the features of a face. The model instance is placed near the centre of the image and a coarse to fine search performed. The search starts at level 3 (one-eighth of the resolution in x and y compared to the original image). Large movements are made in the first few iterations, getting the position and scale roughly correct. As the search progresses to finer resolutions more subtle adjustments are made. The final convergence (after a total of 18 iterations) gives a good match to the target image. In this case at most five iterations were allowed at each resolution, and the algorithm converges in much less than a second (on a 200 MHz PC).

Figure 14 demonstrates how the ASM can fail if the starting position is too

7: Model-based methods in analysis of biomedical images

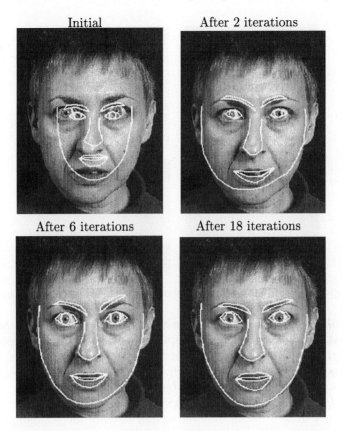

Figure 13. Search using active shape model of a face.

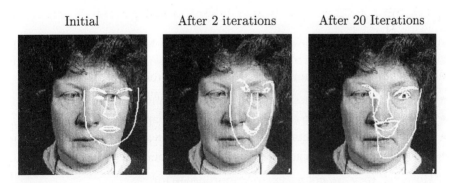

Figure 14. Search using active shape model (ASM) of a face, given a poor starting point. The ASM is a local method, and may fail to locate an acceptable result if initialized too far from the target.

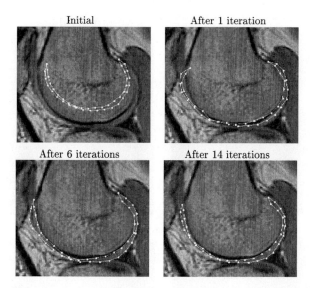

Figure 15. Search using active shape model of cartilage on an MR image of the knee.

far from the target. Since it is only searching along profiles around the current position, it cannot correct for large displacements from the correct position. It will either diverge to infinity or converge to an incorrect solution, doing its best to match the local image data. In the case shown it has been able to locate half the face, but the other side is too far away.

Figure 15 demonstrates using the ASM of the cartilage to locate the structure in a new image. In this case the search starts at level 2, samples at two points either side of the current point and allows at most five iterations per level. A detailed description of the application of such a model is given by Solloway *et al.* (11).

5. Discussion

Active shape models allow rapid location of the boundary of objects with similar shapes to those in a training set, assuming we know roughly where the object is in the image. They are particularly useful for:

- Objects with well-defined shape (e.g. bones, organs, faces, etc.)
- Cases where we wish to classify objects by shape/appearance
- Cases where a representative set of examples is available
- Cases where we have a good guess as to where the target is in the image

However, they are not necessarily appropriate for

- Objects with widely varying shapes (e.g. amorphous things, trees, long wiggly worms, etc.)

7: Model-based methods in analysis of biomedical images

- Problems involving counting large numbers of small things
- Problems in which position/size/orientation of targets is not known approximately (or cannot be easily estimated).

In addition, it should be noted that the accuracy to which they can locate a boundary is constrained by the model. The model can only deform in ways observed in the training set. If the object in an image exhibits a particular type of deformation not present in the training set, the model will not fit to it. This is true of fine deformations as well as coarse ones. For instance, the model will usually constrain boundaries to be smooth, if only smooth examples have been seen. Thus if a boundary in an image is rough, the model will remain smooth, and will not necessarily fit well. However, using enough training examples can usually overcome this problem.

One of the main drawbacks of the approach is the amount of labelled training examples required to build a good model. These can be very time-consuming to generate. However, a 'bootstrap' approach can be adopted. We first annotate a single representative image with landmarks, and build a model from this. This will have a fixed shape, but will be allowed to scale, rotate, and translate. We then use the ASM algorithm to match the model to a new image, and edit the points which do not fit well. We can then build a model from the two labelled examples we have, and use it to locate the points in the third. This process is repeated, incrementally building a model, until the ASM finds new examples sufficiently accurately every time, so needs no more training.

Both the shape models and the search algorithms can be extended to 3D. The landmark points become 3D points, the shape vectors become $3n$-dimensional for n points. Although the statistical machinery is identical, a 3D alignment algorithm must be used (18). Of course, annotating 3D images with landmarks is difficult, and more points are required than for a 2D object. In addition, the definition of surfaces and 3D topology is more complex than that required for 2D boundaries. However, 3D models which represent shape deformation can be successfully used to locate structures in 3D datasets such as MR images (see, for example, ref. 19).

The ASM is well suited to tracking objects through image sequences. In the simplest form the full ASM search can be applied to the first image to locate the target. Assuming the object does not move by large amounts between frames, the shape for one frame can be used as the starting point for the search in the next, and only a few iterations will be required to lock on. More advanced techniques would involve applying a Kalman filter to predict the motion (19, 20).

The shape models described above assume a simple gaussian model for the distribution of the shape parameters, **b**. A more general approach is to use a mixture of gaussians. We need to ensure that the model generates plausible shapes. Using a single gaussian we can simply constrain the parameters with a bounding box or hyper-ellipse. With a mixture of gaussians we must arrange that the probability density for the current set of parameters is above a suitable threshold. Where it is below, we can use gradient ascent to find the nearest point in parameter space which does give a plausible shape (22). However, in practice, unless large non-gaussian shape variations are observed and the target images are noisy or cluttered, using a single gaussian approximation works perfectly well.

To summarize, by training statistical models of shape from sets of labelled examples we can represent both the mean shape of a class of objects and the common modes of shape variation. To locate similar objects in new images we can use the ASM algorithm which, given a reasonable starting point, can match the model to the image very quickly.

6. Implementation

Although the core mathematics of the models described above are relatively simple, a great deal of machinery is required to actually implement a flexible system. This could easily be done by a competent programmer. However, implementations of the software are already available.

The simplest way to experiment is to obtain the MatLab package implementing the ASMs, available from Visual Automation Ltd. This provides an application which allows users to annotate training images, to build models, and to use those models to search new images. In addition, the package allows limited programming via a MatLab interface.

See http://www.isbe.man.ac.uk/VAL for details.

It is intended that a free (C++) software package implementing ASMs will be provided for the Image Understanding Environment (a free computer vision library of software). For more details and the latest status, see http://www.isbe.man.ac.uk/research/IUE/flier.html.

In practice, the algorithms work well on a mid-range PC (200 MHz). Search will usually take less than a second for models containing up to a few hundred points.

Details of other implementations will be posted on http://www.isbe.man.ac.uk.

Appendices

A. Aligning the training set

There is considerable literature (6, 23) on methods of aligning shapes into a common coordinate frame, the most popular approach being Procrustes Analysis (6). This aligns each shape so that the sum of distances of each shape to the mean ($D = \Sigma |\mathbf{x}_i - \bar{\mathbf{x}}|^2$) is minimized. It is poorly defined unless constraints are placed on the alignment of at least one of the shapes.

7: Model-based methods in analysis of biomedical images

Typically, one would ensure the shapes are centred on the origin, have a mean scale of unity and some fixed but arbitrary orientation.

Although analytic solutions exist to the alignment of a set (23), a simple iterative approach is as follows:

Protocol 4. Aligning a set of shapes

1. Translate each example so that its centre of gravity is at the origin.
2. Choose one example as an initial estimate of the mean shape and scale so that

   $$|\bar{\mathbf{x}}| = \sqrt{\bar{x}_1^2 + \bar{y}_1^2 + \bar{x}_2^2 \cdots} = 1.$$

3. Record the first estimate as $\bar{\mathbf{x}}_0$ to define the default orientation.
4. Align all the shapes with the current estimate of the mean shape.
5. Re-estimate the mean from aligned shapes.
6. Apply constraints on scale and orientation to the current estimate of the mean by aligning it with $\bar{\mathbf{x}}_0$ and scaling so that $|\bar{\mathbf{x}}| = 1$.
7. If not converged, return to step 4.

   (Convergence is declared if the estimate of the mean does not change significantly after an iteration.)

The operations allowed during the alignment will affect the shape of the final distribution. A common approach is to centre each shape on the origin, then to rotate and scale each shape into the *tangent space* to the mean so as to minimize D. The tangent space to $\mathbf{x}_t$ is the hyperplane of vectors normal to $\mathbf{x}_t$, passing through $\mathbf{x}_t$, i.e. all the vectors $\mathbf{x}$ such that $(\mathbf{x}_t - \mathbf{x}).\mathbf{x} = 0$, or $\mathbf{x}.\mathbf{x}_t = 1$ if $|\mathbf{x}_t| = 1$.

To align two shapes, $\mathbf{x}_1$ and $\mathbf{x}_2$, each centred on the origin, we choose a scale, s, and rotation, θ, so as to minimize $|T_{s,\theta}(\mathbf{x}_1) - \mathbf{x}_2|^2$, the sum of square distances between points on shape $\mathbf{x}_2$ and those on the scaled and rotated version of shape $\mathbf{x}_1$. Appendix D gives the optimal solution.

The simplest way to find the optimal point in the tangent plane is to first align the current shape, $\mathbf{x}$, with the mean, allowing scaling and rotation, then project into the tangent space by scaling $\mathbf{x}$ by $1/(\mathbf{x}.\bar{\mathbf{x}})$.

Since we normalize the scale and orientation of the mean at each step, the mean of the shapes projected into the tangent space, $\bar{\mathbf{x}}$, may not be equal to the (normalized) vector defining the tangent space, $\mathbf{x}_t$. We must retain $\mathbf{x}_t$ so that when new shapes are studied, they can be projected into the same tangent space as the original data.

Different approaches to alignment can produce different distributions of the aligned shapes. We wish to keep the distribution compact and keep any non-linearities to a minimum, so recommend using the tangent space approach.

B. Principal component analysis

Principal component analysis (PCA) allows us to find the major axes of a cloud of points in a high-dimensional space. This is useful, as we can then approximate the position of any of the points using a small number of parameters. Given a set of vectors $\{x_i\}$, we apply PCA as follows.

Protocol 5. Principal component analysis

1. Compute the mean of the data,

$$\bar{x} = \frac{1}{s}\sum_{i=1}^{s} x_i. \qquad [12]$$

2. Compute the covariance of the data,

$$S = \frac{1}{s-1}\sum_{i=1}^{s}(x_i - \bar{x})(x_i - \bar{x})^T. \qquad [13]$$

3. Compute the eigenvectors, p_i, and corresponding eigenvalues, λ_i, of S (sorted so that $\lambda_i \geq \lambda_{i+1}$). (When there are fewer samples than dimensions in the vectors, there are quick methods of computing these eigenvectors—see Appendix C.)

4. Each eigenvalue gives the variance of the data about the mean in the direction of the corresponding eigenvector. Compute the total variance from

$$V_T = \sum_i \lambda_i.$$

5. Choose the first t largest eigenvalues such that

$$\sum_{i=1}^{t} \lambda_i \geq f_v V_T, \qquad [14]$$

where f_v defines the proportion of the total variation one wishes to explain (for instance, 0.98 for 98%). (See also Section 4.1.6.)

Given the eigenvectors $\{p_i\}$, we can approximate any of the training set, x, using

$$x \approx \bar{x} + Pb, \qquad [15]$$

where $P = (p_1|p_2|\cdots|p_t)$ contains t eigenvectors of the covariance matrix and b is a t-dimensional vector given by

$$b = P^T(x - \bar{x}). \qquad [16]$$

The eigenvectors p_i essentially define a rotated coordinate frame (centred on the mean) in the original $2n$D space. The parameters b are the most significant coordinates of the shapes in this rotated frame.

For instance, *Figure 16* shows the principal axes of a 2D distribution of vectors. In this case any of the points can be approximated by the nearest

7: Model-based methods in analysis of biomedical images

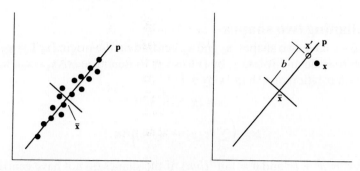

Figure 16. Applying a PCA to a set of 2D vectors. **p** is the principal axis. Any point **x** can be approximated by the nearest point on the line, **x**' (see text).

point on the principal axis through the mean. $\mathbf{x} \approx \mathbf{x}' = \bar{\mathbf{x}} + b\mathbf{p}$ where b is the distance along the axis from the mean of the closest approach to **x**.

C. Applying a PCA when there are fewer samples than dimensions

Suppose we wish to apply a PCA to s nD vectors, $\mathbf{x}_i$, where $s < n$. The covariance matrix is $n \times n$, which may be very large. However, we can calculate its eigenvectors and eigenvalues from a smaller $s \times s$ matrix derived from the data. Because the time taken for an eigenvector decomposition goes as the cube of the size of the matrix, this can give considerable savings.

Subtract the mean from each data vector and put them into the matrix **D**

$$\mathbf{D} = ((\mathbf{x}_1 - \bar{\mathbf{x}})|\cdots|(\mathbf{x}_s - \bar{\mathbf{x}})) \qquad [17]$$

The covariance matrix can be written

$$\mathbf{S} = \frac{1}{s}\mathbf{D}\mathbf{D}^T \qquad [18]$$

Let **T** be the $s \times s$ matrix

$$\mathbf{T} = \frac{1}{s}\mathbf{D}^T\mathbf{D} \qquad [19]$$

Let $\mathbf{e}_i$ be the s eigenvectors of **T** with corresponding eigenvalues λ_i, sorted into descending order. It can be shown that the s vectors $\mathbf{D}\mathbf{e}_i$ are all eigenvectors of **S** with corresponding eigenvalues λ_i, and that all remaining eigenvectors of **S** have zero eigenvalues (24). Note that $\mathbf{D}\mathbf{e}_i$ is not necessarily of unit length so may require to be normalised.

D. Aligning two shapes

Suppose we have two shapes, x_1 and x_2, centred on the origin ($x_1 \cdot 1 = x_2 \cdot 1 = 0$). We wish to scale and rotate x_1 by (s,θ) so as to minimize $|sAx_1 - x_2|$, where A performs a rotation of a shape x by θ. Let

$$a = (x_1 \cdot x_2)/|x_1|^2 \qquad [20]$$

$$b = \left(\sum_{i=1}^{n}(x_{1i}y_{2i} - y_{1i}x_{2i})\right)/|x_1|^2 \qquad [21]$$

Then $s^2 = a^2 + b^2$ and $\theta = \tan^{-1}(b/a)$. If the shapes do not have centroids on the origin, the optimal translation is chosen to match their centroids, the scaling and rotation then chosen as above.

Acknowledgements

Dr Cootes is grateful to the EPSRC for his Advanced Fellowship Grant, and to his colleagues at the Wolfson Image Analysis Unit for their support. The MR cartilage images were provided by Dr C. E. Hutchinson and his team, and were annotated by Dr S. Solloway. The face images were annotated by G. Edwards, Dr A. Lanitis, and other members of the Unit.

References

1. Yuille, A. L., Cohen, D. S., and Hallinan, P. (1992). *International Journal of Computer Vision*, **8**(2), 99–112.
2. Staib, L. H. and Duncan, J. S. (1992). *IEEE Transactions on Pattern Analysis and Machine Intelligence*, **14**(11), 1061–1075.
3. Kass, M., Witkin, A., and Terzopoulos, D. (1987). In *1st International Conference on Computer Vision*, London, pp. 259–268.
4. Grenander, U. and Miller, M. (1993). *Journal of the Royal Statistical Society B*, **56**, 249–603.
5. Mardia, K. V., Kent, J. T., and Walder, A. N. (1991). In *23rd Symposium on the Interface,* Interface Foundation, Fairfax Station, pp. 550–557.
6. Goodall, C. (1991). *Journal of the Royal Statistical Society B*, **53**(2), 285–339.
7. Bookstein, F. L. (1989). *IEEE Transactions on Pattern Analysis and Machine Intelligence*, **11**(6), 567–585.
8. Kirby, M. and Sirovich, L. (1990). *IEEE Transactions on Pattern Analysis and Machine Intelligence*, **12**(1), 103–108.
9. McInerney, T. and Terzopoulos, D. (1996). *Medical Image Analysis*, **1**(2), 91–108.
10. Lanitis, A., Taylor, C., and Cootes, T. (1997). *IEEE Transactions on Pattern Analysis and Machine Intelligence*, **19**(7), 743–756.
11. Solloway, S., Hutchinson, C., Waterton, J., and Taylor, C. (1996). In *4th European Conference on Computer Vision, Cambridge, England* (ed. B. Buxton and R. Cipolla), Vol. 2, pp. 400–411. Springer.
12. Press, W., Teukolsky, S., Vetterling, W., and Flannery, B. (1992). *Numerical recipes in C*, 2nd edition. Cambridge University Press.

13. Goldberg, D. E. (1989). *Genetic algorithms in search, optimisation and machine learning*. Addison-Wesley, Wokingham, UK.
14. Haslam, J., Taylor, C., and Cootes, T. (1994). In *5th British Machine Vision Conference, York, UK* (ed. E. Hancock), BMVA Press, Sheffield, UK, pp. 33–42.
15. Cootes, T., Edwards, G. J., and Taylor, C. J. (1998). In *5th European Conference on Computer Vision* (ed. H. Burkhardt and B. Neumann), Springer, Berlin. Vol. 2, pp. 484–498.
16. Hill, A., Cootes, T. F., and Taylor, C. J. (1992). In *3rd British Machine Vision Conference* (ed. D. Hogg and R. Boyle), Springer-Verlag, London, pp. 276–285.
17. Burt, P. (1984). *The pyramid as a structure for efficient computation*, Springer-Verlag, Berlin, pp. 6–37.
18. Hill, A., Cootes, T. F., and Taylor, C. J. (1996). *Image and Vision Computing*, **14**(8), 601–607.
19. Hill, A., Cootes, T. F., Taylor, C. J., and Lindley, K. (1994). *Journal of Medical Informatics*, **19**(1), 47–59.
20. Baumberg, A. and Hogg, D. (1994). In *3rd European Conference on Computer Vision*, Springer-Verlag, Berlin, **1**, 299–308.
21. Cootes, T. and Taylor, C. (1997). In *8th British Machine Vision Conference, Colchester, UK* (ed. A. Clarke), BMVA Press, Essex, pp. 110–119.
22. Edwards, G. J., Taylor, C. J., and Cootes, T. (1997). In *8th British Machine Vision Conference, Colchester, UK* (ed. A. Clarke), BMVA Press, Essex. pp. 130–139.
23. Dryden, I. and Mardia, K. V. (1998). *The statistical analysis of shape*. Wiley, London.
24. Cootes, T. F., Cooper, D. H., Taylor, C. J. and Graham, J. (1995). *Computer Vision and Image Understanding*, **61**, 38–59.

8

Projective stereology in biological microscopy

ANDREW D. CAROTHERS

1. Introduction

Stereology can be defined as the three-dimensional interpretation of flat images, such as sections and projections, by criteria of geometric probability. Although the definition encompasses both sections and projections, the two approaches differ in both theoretical underpinning and practical application. Most standard theory is concerned almost exclusively with sectioned material. No doubt this is a result of the early historical development of the subject, which was largely driven by geological applications where the main requirement was to determine patterns of inclusions within solid rocks. Sectioning is necessary wherever there is a need to determine the internal composition of large, opaque objects, such as body organs. There is little point in attempting to reproduce here the many useful results from this body of theory. For these, the reader is referred to standard texts (see for example refs 1–3). In many other applications of biological microscopy, however, it is more relevant to consider plane projections since sectioning is invasive and may be technically difficult, and since objects of interest, such as subcellular elements, whole cells, and tissues, are often transparent or nearly so. Another, and perhaps the earliest, application of the stereology of projections was the inference by the ancient Greeks that the earth was spherical from the shape of its shadow on the moon during lunar eclipses.

In this chapter some of the ways in which projections or 'shadows' can be used to make inferences about three-dimensional structures are considered. However, some caution is needed in applying this approach to microscopic biological preparations. The methods discussed in what follows generally require the assumptions first that the objects of interest are projected orthogonally onto the field of view, and second that they are oriented uniformly randomly with respect to the plane of view. Neither assumption is likely to be strictly fulfilled on slide preparations of biological material. The processes of fixing, staining, and mounting cells under coverslips may introduce distortions, violating the first assumption, and sheets of tissue generally contain cells that are all similarly aligned, violating the second assumption. Nevertheless, there are many situations in which the assumptions are a reasonable approximation to reality, and where the ease and efficiency of measuring in two as opposed to three dimensions make it feasible to adopt the approaches discussed here.

2. Points and distances

Suppose a scatter of points is randomly distributed in three dimensions, and that the positions of their orthogonal projections onto a uniformly randomly-oriented plane are recorded. If we represent the three-dimensional distance between each point and some reference point by x, and the two-dimensional distance between their projections by y, what can be inferred about the distribution of x from the observations of y? That the random variables x and y are related is illustrated for one particular situation in *Figure 1*. In fact, standard statistical theory makes it possible to compute all the moments of the distribution of x in terms of those of y (see *Box 1*). From these results, mean(x) can be estimated as $4/\pi$ times mean(y), and its standard deviation as the same multiple of the standard deviation of mean(y).

Box 1. Moments of the 3D random variable, x, in terms of those of its 2D orthogonal projection, y, onto a randomly-oriented plane

If the ith moment of a variable is denoted by $m_i(.)$ (e.g. $m_i(x) = E(x^i)$), then the first four moments of x are given in terms of those of y by:

$$m_1(x) = \frac{4}{\pi} m_1(y) \qquad [1]$$

$$m_2(x) = \frac{3}{2} m_2(y) \qquad [2]$$

$$m_3(x) = \frac{16}{3\pi} m_3(y) \qquad [3]$$

$$m_4(x) = \frac{15}{8} m_4(y) \qquad [4]$$

Hence, the means and variances are related by:

$$mean(x) = \frac{4}{\pi} m_1(y) \qquad [5]$$

$$var(x) = \frac{3}{2} \left\{ var(y) - \left(\frac{1}{\alpha} - 1\right) m_1^2(y) \right\} \qquad [6]$$

where

$$\alpha = 3\pi^2/32 \qquad [7]$$

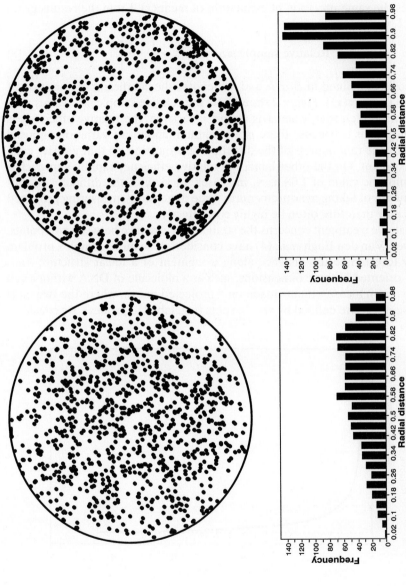

Figure 1. 1000 points uniformly randomly scattered in a sphere of unit radius and projected onto a plane, together with corresponding histograms of radial distances; on the left, points distributed throughout the entire sphere; on the right, points confined to the outer shell of the sphere, more than 0.9 radii from the centre.

It is of interest to compare the efficiency of this method of estimating mean(x) with that of directly measuring the x-values themselves. This amounts to determining the size of a sample of y-values, measured in two dimensions, relative to that of a sample of x-values, measured in three dimensions, which achieves the same precision of estimation of mean(x). From the results given in *Box 1*, we find

$$\text{Relative sample size} = \frac{1}{\alpha}\left\{1 + \frac{1-\alpha}{\gamma_x^2}\right\} \qquad [8]$$

where α is as defined in *Box 1*, and γ_x denotes the coefficient of variation of x ($=$SD(x)/mean(x)). *Figure 2* shows how the relative sample size varies with γ_x. It can be seen that, when x is almost invariable, the relative efficiency is very poor. This is because there is a minimum amount of variation imposed upon y by the randomness of the direction of projection, so that y varies even if x is constant. On the other hand, the relative efficiency rapidly approaches an asymptotic value of 1.08 as γ_x increases. In view of the difficulty and inconvenience of taking measurements in three dimensions, measuring projected images may therefore often be highly cost-effective.

The above treatment concerns the straight-line distance between two points. However, Van den Engh *et al.* (4) have considered the more complex problem of how to estimate the distance along a segment of a linear structure 'randomly' oriented in three dimensions, such as a molecule of DNA within a cell nucleus, from measurements made on a projected image, where the two ends of the segment are defined by visible spots (e.g. labelled probes). By consider-

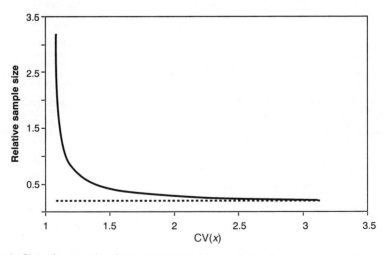

Figure 2. Size of a sample of measurements in two dimensions relative to that in three dimensions required to give the same precision of estimation of the mean distance of a set of random points from a reference point. CV(x) denotes the coefficient of variation of the three-dimensional distances, x.

8: Projective stereology in biological microscopy

ing the linear structure to consist of a large number of small links each of which has freedom to rotate about its point of contact with its neighbour on one side (technically a Gaussian chain), they found that the mean projected distance between the ends of the segment is proportional to the square root of its length. In theory, the constant of proportionality is known, allowing a direct method of estimating the length of the segment. However, because this is a model-based approach involving untested assumptions, Van den Engh et al. adopted the more cautious policy of *calibrating* the relationship by measuring the projected distances between DNA probes of known location, and were thereby able to show that the linearity of the relationship held for genomic distances between 100 and 1500 kbp. Where the relationship holds, it has been shown that, in order to estimate the interprobe distance with coefficient of variation, γ, it is necessary to measure γ^{-2} spot-pairs (5). Thus, to achieve coefficients of variation of 10% and 1% by this method requires measurements on 100 and 10000 spot-pairs respectively.

3. Lines

A technique for estimating the length of a general curve in two dimensions was invented as long ago as 1777, and has come to be known as 'Buffon's needle' (6). *Protocol 1* explains the method, and how it can be applied to the case of a general curve in three dimensions projected onto a randomly oriented plane.

Protocol 1. Estimating the length of a plane curve (Buffon's needle)

Method

1. Place or project the curve to be measured randomly onto a grid of parallel lines with adjacent lines separated by a distance, *d* (see *Figure 3*). In practice, it may be more convenient to place the grid randomly over the curve or its projected image, but this has no effect on the method of estimation.

2. Count the number of intersections, N_i, between the curve and the lines of the grid.

3. Estimate the length, L, of the curve by $\hat{L} = \beta d N_i$, where β takes the value $\pi/2$ when the curve is two-dimensional and in the plane of projection, and the value 2 when it is randomly oriented in three dimensions.

Unfortunately, there is no general formula for the *variance* of the estimated length ($\hat{L}$), since this depends on the shape of the curve. For example, in the

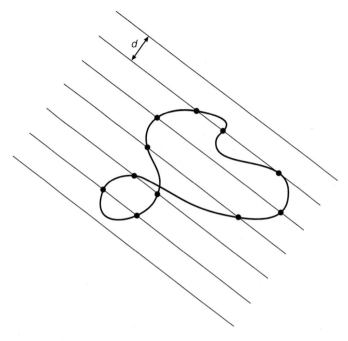

Figure 3. Buffon's needle. A method for estimating the length of a curve.

case of a circle within the plane of projection the coefficient of variation of $\hat{L}$ lies between 0 and $\pi d/2L$, and can therefore be made arbitrarily small by choosing d sufficiently small relative to L, or equivalently by increasing the magnification. For a straight line, however, the coefficient of variation is approximately 0.48, irrespective of the spacing of the grid, provided that $d \ll L$. There is therefore a limit to how far one can improve the precision of estimation by reducing d. With a sample of m independent measurements, the coefficient of variation of the mean is equal to $CV(\hat{L})/\sqrt{m}$ and can be made arbitrarily small by choosing a sufficiently large m.

When a curve can be viewed in plane projection from different angles that can be precisely controlled by the observer, then the method of total vertical projection (TVP) can be used (7). However, since there is no simple way to determine the precision of the resulting estimate, it is difficult to compare its efficiency with that of an estimate based on a simple Buffon's needle approach.

4. Surface areas

The area of a circle of radius r is πr^2, while the surface area of a sphere of radius r is $4\pi r^2$. Thus if we denote the surface area by A_S, and the projected area by A_P, then $A_S = 4A_P$. Surprisingly, this simple relation holds for any convex object, irrespective of shape, provided that A_P is understood as the *mean projected area* onto uniformly randomly oriented planes (8). (An object is convex if and only if the straight line joining any two points within it, or on its surface, is wholly contained within it). This provides a simple way to obtain an unbiased estimate of the surface area of any such object (*Protocol 2*).

Protocol 2. Estimating the surface area of a convex object

Method

1. Project the object onto a randomly oriented plane.
2. Measure the projected area, $\hat{A}_P$.
3. Estimate the surface area by $4\hat{A}_P$.
4. To improve precision, repeat the process on several random planes and take the mean value.

As for the estimation of curve length by Buffon's needle, there is no simple relation that makes it possible to determine the *variability* of this estimate, since it depends upon the shape of the object. To illustrate this, we consider the simplest possible generalization of a sphere, namely an ellipsoid in which two axes are identical, of length a say, and one is different, of length c say. Such an object is egg-shaped, spherical, or disc-shaped depending on whether the ratio t (= a/c) is less than, equal to, or greater than one, respectively. Standard geometrical and statistical arguments make it possible to compute the distributional properties of the ratio $R_S = \hat{A}_S/A_S$, where $\hat{A}_S$ denotes the estimated surface area (= $4\hat{A}_P$). In particular, the mean and standard deviation of R_S are shown in *Figure 4*. The relation $A_S = 4A_P$ implies that the mean is one irrespective of the value of the axis ratio t, but the standard deviation increases towards an asymptotic value of 0.58 as $t \to \infty$, and of 0.28 as $t \to 0$. Hence an upper limit on the standard error of the mean of m independent measurements is $0.58/\sqrt{m}$. It is tempting to conjecture that this upper limit is applicable to convex objects of any shape, but I know of no proof of this. Note that when $t = 1$ the standard deviation is zero, since a sphere projects the same area in all orientations. There is of course no way of dealing, even in principle, with non-convex objects, since holes, indentations, etc. may be hidden from view in all orientations.

5. Volumes

The volume, V, of a sphere of radius r is $4\pi r^2/3$, and hence could be estimated from its projected area, A_P, as:

$$\hat{V} = \frac{4}{3\sqrt{\pi}} (A_P)^{3/2} \qquad [9]$$

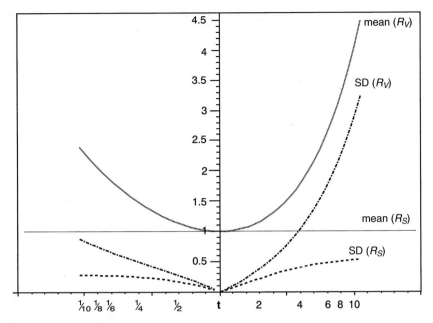

Figure 4. Means and standard deviations (SD) of estimates of the surface area (R_S) and volume (R_V) of an oblate ($t < 1$) or a prolate ($t > 1$) ellipsoid, relative to their true values, from measurements of the projected area onto a uniformly random plane. Parameter t denotes the ratio of the length of the two common axes to that of the different one.

In contrast to the estimation of surface area, this does not give unbiased estimates for a wider class of objects. Nevertheless, it is of some interest to consider how well it performs for the class of ellipsoids described above. *Figure 4* shows the dependence on t of the mean and standard deviation of the ratio $R_V = \hat{V}/V$. It can be seen that the mean is always positively biased, and that both mean and standard deviation increase, apparently without limit, as t departs from one. However, provided that $0.6 < t < 1.5$, the bias is less than 5% and can probably be ignored for most practical purposes. For comparison, a typical chicken's egg has a t-value of about 0.75.

6. Applications

6.1 Determining the order of three DNA loci

The method of Van den Engh *et al.* (4) described above can be used to determine the order of three closely linked DNA loci, by hybridizing different coloured probes to them and then examining a large number of such triplets to see which pair of colours is separated most widely in plane projection. Denoting the ith locus by L_i, the correct order by L_1–L_2–L_3, and the correct

8: Projective stereology in biological microscopy

interlocus distances by d_{12}, d_{13}, and d_{23} (by definition therefore $d_{13} = d_{12} + d_{23}$), we may ask how many triplets need to be measured in order to infer the correct order with a given level of confidence, $1 - \alpha$ say. In the most difficult situation where two loci are situated close together relative to the third (that is, when for example $d_{13} \approx d_{12}$, or equivalently when $d_{23} \ll d_{12}$), it has been shown that the required number is approximately

$$k_\alpha^2 \left\{ 1 + \frac{2d_{12}}{d_{13}} \right\} \quad [10]$$

where k_α is the standard normal deviate corresponding to a tail area of α (5). For example, if $d_{12} = 10d_{23}$, and we wish to reduce the probability of inferring the wrong order to 0.01, then $k_\alpha = 2.33$ and the required number is about 114. If $d_{12} = 100d_{23}$, then the number increases to 1090.

6.2 Location of chromosomal domains in cell nuclei

Suppose a sphere of unit radius is viewed in plane projection. The proportion of the sphere's volume that is contained within a radius, s, of the projected circle is, from basic geometry,

$$1 - (1 - s^2)^{3/2} \quad [11]$$

This simple result can be used to test the null hypothesis that subcellular structures are uniformly distributed throughout the nucleus. For example, in an experiment to study the location of chromosome 19 in interphase human primary lymphocytes, each projected nucleus was divided into five concentric bands of equal area, and the amount of chromosome-specific stain was measured within each band (9). As a control, the nucleus was counterstained with a non-chromosome-specific DNA stain (DAPI). The procedure is illustrated in *Figure 5*, and results are summarized in *Table 1*. It can be seen (i) that the amount of DAPI stain agrees closely with prediction, suggesting

Table 1. Ratio of observed to expected stain in interphase primary lymphocytes

Radius of band	Volume (% of total)	DAPI stain (O/E)	Chromosome-19 stain (O/E)
0.89–1.00	8.9	1.3	0.1
0.77–0.89	16.4	1.0	0.3
0.63–0.77	21.2	1.0	0.7
0.45–0.63	25.1	1.0	1.3
0.00–0.45	28.1	0.9	1.6

Cell nuclei are standardized to have a radius of 1. Column 1 shows the inner and outer radii of successive bands on the projected image. Column 2 shows the theoretical proportion of total volume projected onto each band. Columns 3 and 4 show the ratio of observed to expected stain in each band, where the expected value is proportional to the corresponding volume (reproduced from Table 6.8 of ref. 9).

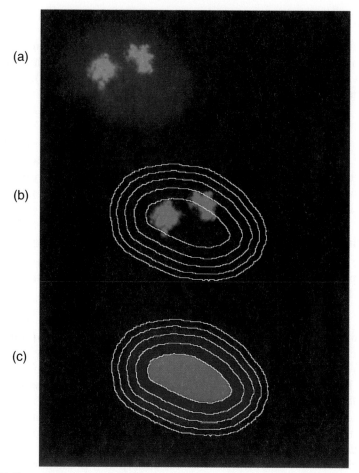

Figure 5. Determining the distribution of FISH signal within ellipsoid nuclei. (a) Chromosome 19 FISH paint labelled with biotin and detected with avidin-FITC (green); nucleus counterstained with DAPI. The nucleus was automatically segmented from the background and the area and centroid coordinates were calculated. The area was divided into five equal segments by repeated erosion of the outer boundary from the periphery towards the centre of the nucleus. For each segment the amounts of FISH (b) and DAPI (c) fluorescence were calculated and represented as a percentage of the total.

that total DNA is indeed evenly distributed throughout the nucleus, and (ii) that the amount of chromosome 19-specific stain differs markedly from that predicted by the null hypothesis, and suggests that chromosome 19 is generally located near the nuclear centre in these cells. The mean distance of chromosome 19 from the nuclear centre in 3D could of course be calculated from its mean distance in 2D by using the multiplier $4/\pi$, as explained in *Box 1*.

6.3 Size of chromosomal domains in cell nuclei

If the proportion of projected area occupied by a particular chromosome is f, then the proportion of total volume occupied by the chromosome is approximately $f^{3/2}$, assuming that both the nucleus and the chromosomal domain are approximately spherical. For example, as part of the study quoted in the previous section, Croft (9) showed that the proportions of projected nuclear area occupied by chromosome 19-specific stain during G1, early S, late S, and G2 phases were respectively 7.3, 9.3, 7.7, and 6.6%. Hence the corresponding proportions of total volumes could be estimated as respectively 2.0, 2.8, 2.1, and 1.7%, whereas the DNA content of chromosome 19 represents approximately 2.6% of the total genome. Comparisons of these volume proportions at different stages of the cell cycle, or with other chromosomes, may provide useful clues to chromosomal functioning and/or replication timing, and are generally more meaningful than the measured area proportions.

7. Discussion

The applicability of the rather simple-minded approaches described here is limited by the need to satisfy certain assumptions, as noted in the Introduction. Where the assumptions are seriously violated, more elaborate techniques may be necessary, usually involving physical or optical sectioning or strict control over the angle of projection of individual objects (see for example ref. 10 for a review). In view of the sensitivity to assumptions, it is strongly recommended that wherever possible these methods should be *calibrated* by applying them to test systems with known properties, before using them routinely on a large scale. For example, it may be feasible to compare measurements from two-dimensional cell preparations with those based on confocal microscopy of cells in a more 'natural' three-dimensional configuration. Alternatively, as in the study of Van den Engh *et al.* (4), the test system may involve known physical distances. In some cases, although the simple approach may suggest the general form of a relationship between the two-dimensional and three-dimensional properties of a population of objects, the exact parametric relationship may require experimental determination. An example concerns the determination of the length of a curve using Buffon's needle. As we saw in *Protocol 2*, the multiplier β takes the values $\pi/2$ or 2 depending on whether the original curve is two-dimensional and in the plane of projection (as might be the case for a relatively large structure within a squashed cell mounted under a coverslip) or randomly oriented in three dimensions (as might be the case for a smaller structure). In reality, the linear relationship might still be valid but with a multiplier of intermediate value.

An entirely different approach that may be applicable in certain situations where the theory is intractable is to use Monte Carlo simulation. Schleicher *et al.* (11) adopted this approach in a study of the distribution of nucleoli within cell nuclei. Because the volumes of individual nucleoli were large relative to those of entire nuclei, it was not possible to treat them as idealized points. Simulation has the further advantage of providing an estimate of the entire distribution of the quantities being evaluated rather than just the means, so that conventional significance tests can be applied.

Acknowledgements

I am indebted to Norman Davidson for the artwork, and to Jenny Croft for permission to reproduce data and *Figure 5* from her PhD thesis.

References

1. Jensen, E. B. (1997). *Local stereology.* World Scientific. Singapore, London.
2. Elias, H. and Hyde, D. M. (1983). *A guide to practical stereology.* Karger, Basel.
3. Howard, C. V. and Reed, M. G. (1998). *Unbiased stereology.* BIOS Scientific, Oxford.
4. Van den Engh, G., Sachs, R., and Trask, B. J. (1992). *Science,* **257**, 1410.
5. Carothers, A. D. (1994). *Cytometry,* **16**, 298.
6. Buffon, G. L. L. (1777). *Supplément à l'histoire naturelle,* Vol. 4. Imprimerie Royale, Paris.
7. Cruz-Orive, L. M. and Howard, C. V. (1991). *Journal of Microscopy,* **163**, 101.
8. Underwood, E. E. (1970). *Quantitative stereology.* Addison-Wesley, Massachusetts.
9. Croft, J. A. (1998). *Correlating mammalian chromosome structure and function.* PhD thesis, University of Edinburgh.
10. Cruz-Orive, L. M. and Weibel, E. R. (1990). *American Journal of Physiology,* **258**, 148.
11. Schleicher, A., Zilles, K., and Kretschmann, H.-J. (1980). *Mikroskopie,* **37**(suppl.), 225.

9

Image warping and spatial data mapping

RICHARD A. BALDOCK and BILL HILL

1. Introduction

Modifying the geometry of (*warping* or *morphing*) a digital image is important for many purposes. D'Arcy Thompson (1) used coordinate transformations to model variations in form with a view to understanding the underlying processes controlling growth. More recently, warping has been required as part of the data capture or reconstruction process, i.e. to restore what is assumed to be the correct geometry. Examples are to correct for spatial distortion in the imaging process, such as perspective transformation or spherical aberration in the optics, or to convert from one coordinate frame to a rectilinear array such as in ultrasound scanning or computed tomography, or to correct for geometric distortions introduced in the material processing before imaging such as in microtome sections of tissue. In medical imaging complex distortion of the geometry of images may be necessary to compare images from one scanning modality to another (see Chapter 6), to compare images from the same individual over time, to average between individuals, and to compare an individual with an atlas. Morphing has also been used for special effects, particularly in film and television and for simulating morphogenesis (2). With the development of spatially organized databases, e.g. geographic information systems (GISs) or, more recently, biological databases such as the Brain Map project in the USA (3) or the Edinburgh Mouse Atlas and gene-expression database (4), it is important that data accumulated at different times, with different instruments, and in the biological situation from different individuals, are mapped onto the same standard coordinate frame.

In the case of a GIS the standard frame is typically part of the earth's surface, perhaps with depth, and all measurements have been made on the same 'ground', and the required warp is to correct straightforward global geometric distortions such as perspective or scaling. In the biological case the standard frame may be an idealized atlas or a representative individual deemed to be 'typical'. In the biological case the data will arise from many different individuals for which, natural variation, processing distortions and different imaging techniques will result in distortions from the standard that include a large random component. In this chapter we discuss techniques for the correction of such distortions in the context of mapping data from one individual onto a standard model. For our application the requirement is to map from a given 2D image or source image onto a 2D destination image (usually a section or a projection from a 3D image).

For this purpose the process of image warping can be considered in three parts. The first is to define the correspondences between the source and destination images. These are pairs of coordinates, one in each image, that are known to match, so that after the warp the coordinate in the source image will equal the corresponding coordinate in the destination image. These are also termed 'tie-points'. The correspondence or match between two images could also be defined by other image structures, for example lines or curves or regions in which the actual coordinate transform is implicit (see also Chapter 5).

The second part is to define the transformation for each pixel in the image given the transformation at a number of given locations. There has to be a mechanism to interpolate from the known transformations at the tie-points to the rest of the 2D space. These transforms can be classed as *global*, in which the interpolating functions extend over all space, or *local*, in which the functions extend piecewise over limited domains. The interpolating functions may be parameterized splines or derived by considering a physical model of deformation, i.e. *model-based*.

The third component of image warping is the process of scanning the source or destination image transforming the pixel values as required. This process implies that the warp transform is calculated for all pixels in the destination image and the pixel value estimated by reference to the source image. The calculation of the transform for each pixel location can be very slow, but for the typical warp of biological data it can be approximated by dividing the image into small regions (elements), typically triangles, within which the transformation is assumed to be affine and therefore calculated much more quickly.

1.1 Notation and numerical methods

In this chapter we use the notation I or I_k to represent an image which is composed of a number of pixels p_i. These pixels, which are possibly scalar values, typically lie on a rectilinear coordinate grid. Each pixel is usually considered to be a digitized *sample* of an underlying 'real' image which is assumed to be continuous in both its spatial and intensity dimensions. The imaging process determines how each pixel corresponds to the original image value in its neighbourhood. Modern CCD array cameras (see Chapter 1), which are the most convenient choice for most microscope applications, are capable of providing a linear response to the incident light intensity and the better cameras have a square sampling grid.

9: Image warping and spatial data mapping

For a given image the pixels are located within an region of 2D integer space defined by a point-set and termed a *domain*, D_k, which in many cases will be rectangular and accessed raster-fashion. Thus

$$I_k = \{p_i^k : i \in D_k\}. \quad [1]$$

The spatial coordinates of each pixel $r_i = (x_i, y_i)$ are an integer-value pair; in some cases we will use the shorthand notation p_{ij} for the pixel value p_k when the spatial coordinates are $r_k = (i, j)$. For any finite image $i_{\min} \leq i \leq i_{\max}$ and $j_{\min} \leq j \leq j_{\max}$, the pair of coordinates $r_{\min} = (i_{\min}, j_{\min})$ and $r_{\max} = (i_{\max}, j_{\max})$ defines the *bounding-box* of the image domain. The pixel values could be a single greyvalue or a vector of values, commonly colour, represented as three-channel red–green–blue:

$$p_k = \begin{cases} b_i & \text{—binary values} \\ g_i & \text{—grey values} \\ (r_i, g_i, b_i) & \text{—colour values} \end{cases} \quad [2]$$

In principle, an image could have any number of channels; for example, LandSat satellite images of the earth are seven-channel images, with each channel corresponding to sensors responding to different wavelengths.

The 2D warp transformations considered here can be written as a mapping from coordinate $\mathbf{r} = (x, y)$ to $\mathbf{w} = (u(x, y), v(x, y))$ with image $I(x, y)$ transformed to $I'(u, v)$. In most of the warp transforms presented in this chapter

$$u(x, y) = \sum_{i=0}^{N} \lambda_i f_i(x, y),$$
$$v(x, y) = \sum_{i=0}^{N} \mu_i g_i(x, y), \quad [3]$$

where $f_i(x, y)$ and $g_i(x, y)$ are the warp basis functions defined below for each transform. The specific transformation is therefore determined by the coefficients λ_i and μ_i which are estimated by solving the system of simultaneous algebraic equations given by substituting the M tie-point values into *Equation 3*. The set of simultaneous equations can be cast into the matrix form

$$\begin{pmatrix} f_0(x_1, y_1) & f_1(x_1, y_1) & \cdots & f_N(x_1, y_1) \\ f_0(x_2, y_2) & f_1(x_2, y_2) & \cdots & f_N(x_2, y_2) \\ \vdots & \vdots & & \vdots \\ f_0(x_M, y_M) & f_1(x_M, y_M) & \cdots & f_N(x_M, y_M) \end{pmatrix} \begin{pmatrix} \lambda_1 \\ \lambda_2 \\ \vdots \\ \lambda_N \end{pmatrix} = \begin{pmatrix} u_1 \\ u_2 \\ \vdots \\ u_M \end{pmatrix} \quad [4]$$

which is termed the *design* equation for λ_i with a similar equation for μ_i. These simultaneous equations are solved using standard linear algebra techniques. We have used singular value decomposition (5) to solve the design equations because it results in a solution which minimizes the sum of squared distance error for each tie-point for overspecified systems ($M > N + 1$) and underspecified systems ($M \leq N$).

2. Image comparison

For the purpose of interactive mapping of data from one biological source onto a standard reference it is important to be able to review the result of the transformation to assess if more matching information, for example additional tie-points, is required. In the case of matching two greylevel images it can be quite difficult to perceive differences if the images are presented side by side and some mechanism is required to display the structural information of one image overlaid on the other. This can be achieved in a number of ways; the most appropriate will depend on the nature of the image data as well as the computing hardware available.

If the images include strongly defined lines and edges with clear angular structures, then an appropriate method could be to extract those features and show the comparison as a set of overlaid line segments. However, this is rarely the case with biological data; therefore, methods which allow the overlay and comparison of all the image data are preferred. Here we present four techniques. The first two are based on representing one image as a binary overlay, calculated by dithering or by feature extraction, over the second image which can be 'seen-through' the overlay. The third is to use colour to highlight areas where the images mismatch with normal greyvalues displayed where the images match. All of these methods are only applicable to greylevel images or separate colour components. The fourth method is known as a 'blink comparator' and can be used for full-colour as well as greylevel images.

2.1 Binary image overlay

With this method for image comparison, one of the images is converted to a simple binary overlay, by for example dithering (6) or extracting salient features, and overlaid on the other image. The dithered image is a series of dots, closer where the original greyvalues are darker and further apart in light regions as shown in *Figure 1a*. *Figure 1b* shows the result of extracting the most significant edges from the image (see Chapter 2) and using the pixels with high edge strength as the binary mask. This first image can be efficiently overlaid on the greylevel or colour second image by letting it use one bit-plane of the displayed image. This is achieved by summing the two images to form a single composite which can then be displayed or by using the standard graphical interface option of writing to selected bit-planes of the display. The colour look-up table (LUT) of the display can then be manipulated so that the pixels which are part of the binary overlay can be a single block colour (say red) and the second image is left as in its original form. The effect is that the second image can be seen through a mask of the structure of the first image. When the images are aligned, the mask visually 'clicks' into place; edges and regions that do not match are usually easily visible. The operations required for this type of display are dithering, bit-plane setting within the composite image or display, and setting the LUT:

9: Image warping and spatial data mapping

(A)
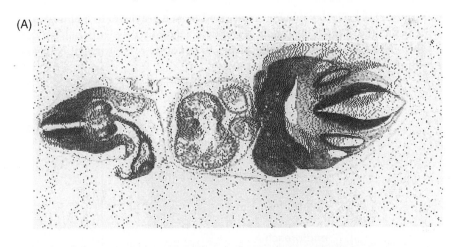

(B)
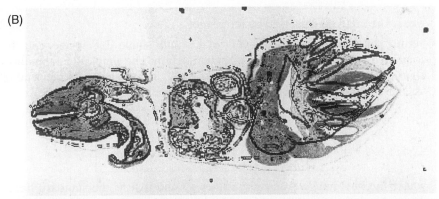

(C)
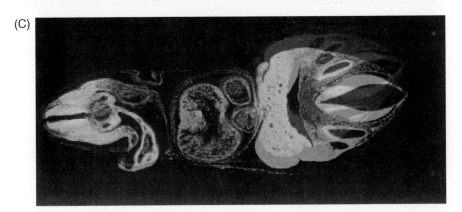

Figure 1. Comparison of two images by binary and colour overlays: (a) dithered overlay; (b) thresholded edge enhanced overlay; (c) colour comparator.

- Dithering: the overlay image is converted from a greylevel to a binary image such that the density of pixels set to 1 is related to the original greylevel value. The Floyd–Steinberg algorithm to generate a 1-bit image I_2 from an 8-bit image I_1 is:

```
set accumulator = 0
for each i ∈ D₁
    accumulator = accumulator + gᵢ¹
    if accumulator > 127
        bᵢ² = 1
        accumulator = accumulator − 128
    else
        bᵢ² = 0
    endif
endforeach
```

where the image is traversed raster fashion.

- Setting bit-planes: the greylevel image and single bit image can be combined before display or as a display function. The simplest combination algorithm uses the operations of bitwise AND, for which we use the symbol '&', and bitwise OR, for which we use the symbol '|' (vertical bar):

```
for each i ∈ D₁
    gᵢ³ = gᵢ¹ & (bᵢ² | 0xfe)
endforeach
```

where the new image I_3 has greyvalues g_i^3 and 0xfe is the standard hexadecimal notation for the number 254, i.e. an 8-bit number with all but the first bit set to 1.

- Defining the LUT: for the algorithm above in which bit-plane 1 corresponds to the dithered overlay. An LUT for which high values result in white and the overlay is red is given by:

```
i = 0
while i < 256
    if i & 1
        red(i) = 255
        green(i) = blue(i) = 0
    else
        red(i) = green(i) = blue(i) = i
    endif
    i = i + 1
endwhile
```

9: Image warping and spatial data mapping

With these algorithms the procedure for displaying the two images is given in *Protocol 1* and illustrated in *Figure 1a*.

Protocol 1. Displaying two images I_1 and I_2 using a dithered overlay

Algorithms
- Floyd–Steinberg dithering to generate a 1-bit image
- Setting a specific bit-plane within the image
- Image processing sequence to extract significant features, e.g. edge filter plus threshold
- Calculating an appropriate LUT

Method
1. Select method for calculating the binary image.
2. Calculate image I_1' as a 1-bit dithered version of I_1 or by feature extraction.
3. Combine images I_1' and I_2 to image I_3 by setting bit-plane 1 to the values of I_1'.
4. Set the display LUT so that odd image values are blank red.
5. Display image I_3.

2.2 Colour comparator

The colour comparator can be used for comparing two greylevel images. The basic idea is to merge the greylevel information and set the LUT so that where the two images are close in value, the colour value displayed is close to the original grey shade and where the images are different the display will be one of two colours (e.g. red/green) depending on which of the two images is dominant. Similar effects can be achieved using the notion of transparency to provide a measure of blending, see for example the OpenGL manual (7). A simple algorithm to combine the greyvalues is to display the colour image with pixel values defined by:

$$p_i = \left(g_{1i}, g_{2i}, \frac{g_{1i} + g_{2i}}{2}\right) \qquad [5]$$

where g_{1i}, g_{2i} are the greyvalues of the original images. If the original greyvalue images had 8-bit values, i.e. $0 \leq g_{1i}, g_{2i} \leq 255$, then the composite colour image requires 256×256 colour values. These may not be available on many machines; therefore, *Equation 5* can be modified to produce a greylevel image

$$g_i = \left(\frac{g_{1i}}{16}\right) \times 16 + \frac{g_{2i}}{16} \qquad [6]$$

where integer division is implied (i.e. round *down* to the nearest integer after division) and the resulting image displayed using the LUT:

```
i = 0, j = 0, k = 0
while i < 16
    while j < 16
        red(k) = 16i
        green(k) = 16j
        blue(k) = 8j(i + j)
        j = j + 1, k = k + 1
    endwhile
    i = i + 1
endwhile
```

See *Figure 1c* for the result of using this LUT for two 8-bit greylevel images. Clearly, the RGB values in the LUT can be adjusted if other colours are preferred in the comparator. Changing one of the images means that the composite image needs to be recalculated because the sum operation for the blue component cannot be achieved by simple masking. This can make the re-display of the combined image slow but, in general, the time to scan and merge two images is much less than the time to calculate the warp transformation.

Protocol 2. Displaying two images I_1 and I_2 using a colour comparator

Algorithms

- Combining two images to a 24-bit RGB image—Equation 5
- Combining two images to an 8-bit image—Equation 6
- Setting an 8-bit LUT—algorithm above

Method

1. Select the display mode—full colour or index (LUT).
2. If full colour mode then combine images to a 24-bit colour image.
3. If index mode then combine the images to an 8-bit image and set the LUT.
4. Display the combined image.

2.3 Blink comparator

The blink comparator (8) relies on the ability of the human visual system to very rapidly make correspondences between two images that are close in terms of content. The primary uses for this ability are motion detection and stereo vision. If two images are nearly the same then the differences will appear as movement of the image structures or as stereo disparities and will be interpreted as depth information. In the case of image comparison and alignment, the differences between the two images may be random in direction and therefore the appropriate use of this ability is to present the images in a time sequence so that the discrepancies appear as movement of the image structures. The alternative of presenting the images as a stereo pair would only be appropriate if all the required adjustments were approximately horizontal, i.e. parallel to the line between the stereo viewpoints.

Two aspects are important if the blink comparator is to work; the first is that the image refresh should be as fast as possible and the second is to ensure that the switch time from one image to the other is regular and constant, i.e. the switching should be at a constant rate. The best way to achieve the first is to use any available hardware support. On many systems it is possible to use 'double buffering'. With double buffering the non-visible image is loaded into video memory and when signalled the display hardware will display that image for the next refresh cycle. This makes the switch from one image to the other instant, i.e. at the refresh rate. The old image can then be modified in the other buffer and when ready the view switched back.

3. Image re-sampling

Once the coordinate transformation from one image to another has been established, for image mapping we need to generate a transformed image with the greylevel information of one image in the coordinate frame of the other. The simplest mechanism for this is to scan the destination image pixels and for each pixel location calculate the reverse transformation to find the corresponding point on the source image. In general, this will not lie on an exact pixel location so it is necessary to obtain an estimate of the greyvalue at the off-pixel coordinate. This is termed image re-sampling and the reliability of the process depends on whether the original signal was sufficiently sampled (see Chapter 1). In this case the correct re-sampling function uses the Sinc interpolation function, strictly over an infinite mask (9). For most situations a simpler (and computationally cheaper) form of estimation is sufficient and for our purposes we have used *nearest neighbour* and *bilinear* interpolation. Any form of interpolation can be considered as a modelling process: an assumption is made about the general shape of the greylevel surface and the idea is to establish a continuous form of that surface to establish the value at an off-pixel coordinate. The quality of the re-sampling will depend on the complexity of the interpolation function but, in general, more complex methods require much longer computation times (see ref. 9 for details of spline and window interpolation).

3.1 Nearest neighbour interpolation

This is the simplest (and quickest) method for estimating the pixel value given a non-integer location and is simply to use the value at the nearest integer location. The algorithm is trivial but care must be taken if the usual round-down operation for floating point to integer conversion is used to make sure that pixels with negative coordinate values are treated correctly. If the result of the inverse map from point $w = (u, v)$ is the coordinate $z = (x, y)$ in the source image then the required pixel value is p_{ij} where

$$i = \begin{cases} \text{Int}(x + 0.5) & \text{for } x \geq 0, \\ \text{Int}(x - 0.5) & \text{for } x < 0 \end{cases} \quad j = \begin{cases} \text{Int}(y + 0.5) & \text{for } y \geq 0, \\ \text{Int}(y - 0.5) & \text{for } y < 0 \end{cases} \quad [7]$$

and $\text{Int}(x)$ is the function converting a float value to an integer (usually invoked implicitly when a floating point value is assigned to an integer variable).

3.2 Bilinear interpolation

Bilinear interpolation assumes that the pixel value between the integer sampling points is linearly related to the distance from the nearest sample points which surround $z = (x, y)$. If these sample points have coordinates given by the four corners of the bounding box $z_{\min} = (x_{\min}, y_{\min})$, $z_{\max} = (x_{\max}, y_{\max})$ then

$$p(x, y) = \frac{y^+ x^+ p(x_{\min}, y_{\min}) + y^+ x^- p(x_{\min}, y_{\min}) + y^- x^+ p(x_{\min}, y_{\min}) + y^- x^- p(x_{\min}, y_{\min})}{(x_{\max} - y_{\min})(x_{\max} - y_{\min})} \quad [8]$$

where $x^+ = (x_{\max} - x)$, $x^- = (x - x_{\min})$, $y^+ = (y_{\max} - y)$, and $y^- = (y - y_{\min})$.

3.3 Computational efficiency

Certain floating point operations take far longer to execute than others. On a Sun Ultra10/300, which was used for all timings, evaluating a four-term polynomial (three multiplications and four additions) takes ~150 ns, evaluating a single logarithm takes ~550 ns. So, for example, a lower bound on a numerically intensive calculation such as the thin-plate splines (section 5.4) is given by:

$$\text{min time} = 2 \times T \times N \times P \quad [9]$$

where T is the time taken to compute a single logarithm, N is the number of tie-points, and P is the number of pixels to be transformed. Given a 512 × 512 image and 20 tie-points, this evaluates to about 5.5 s. With far more tie-points and larger images, as is often required, the evaluation time can become excessive. Experience with mapping gene-expression patterns has shown that for manual warping to be successful, it should be interactive.

Powell (10) has developed an algorithm based on tabulated finite differences for rapidly evaluating thin-plate splines. This has been used by Barrodale et al. (11), who report an improvement of two orders of magnitude in execution time. We have achieved similar results with all radial basis functions using a

9: Image warping and spatial data mapping

less complex algorithm in which we define a triangulated mesh covering the image. The basic idea is to divide the image up into small regions within which it is a good approximation to treat the warp transformation as a simple linear transform (affine). The algorithm then calculates the displacement of each vertex of the mesh using the full warp transform, and within each triangle, or *element*, the image is transformed assuming an affine transform. This means that the complex transform is only calculated at a small number of points, typically two to three orders of magnitude less than the total number of pixels, and the remainder can be calculated using the linear affine transform.

The triangle element size is determined so that the piecewise linear transformation is a good approximation to the original warp transform. This will be application dependent and should be determined by an iterative process, adaptively refining the mesh until the required accuracy is achieved. The method for transforming the whole image is given in *Protocol 3* and illustrated in *Figure 2*.

Protocol 3. Fast image transformation using a triangular mesh

Algorithms

- Triangulation of an image domain
- Interpolation method (see below)
- Computation of an affine transform from three vertex displacements (see Chapter 5)
- Computation of an eight-connected boundary
- Calculation of the intersection of a scan-line with a mesh element

Method

1. Generate a mesh which covers image I with triangular elements, with a triangle size selected so that the linear affine within each element is a good approximation to the required warp.
2. Compute the displacement of each mesh node using the selected warp transformation.
3. For each triangle in the mesh calculate the affine transform which corresponds to the displacement of the three vertices.
4. Compute the eight-connected image boundary B.
5. Compute the transformed image boundary B' by displacing each vertex along B. This is achieved by scanning the mesh to determine which element contains the given boundary pixel and then applying the corresponding affine transform.
6. Create the domain of image I' from the transformed boundary B'.
7. Scan the domain of image I'; for each scan-line compute the intersecting segments with the displaced mesh and hence the affine transform of each segment. Within each segment compute the corresponding location of each pixel in I and estimate the pixel value using the selected grey level interpolation method.

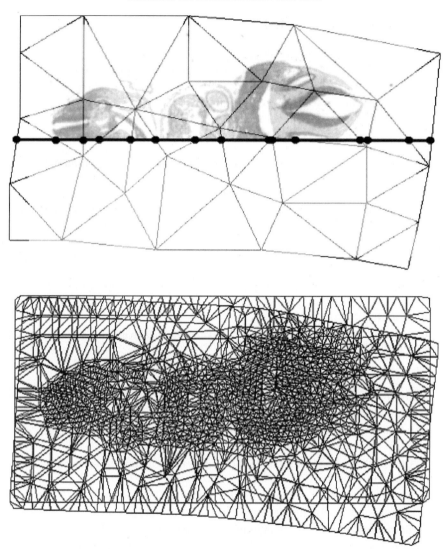

Figure 2. Scanning the warped mesh by traversing the scan-line shown as a dark horizontal line. The scan-line is progressively passed over the mesh from top to bottom. For each position the intersection with each mesh edge is detected (dark dots) and used to segment the scan-line into segments within a single triangle. Each segment maps onto a straight line in the original image via an affine transform and hence the pixel value extracted. As the scan-line passes down the mesh the transformed image is determined (shown above the line). The lower picture shows the actual mesh used in this example, before and after warping. Note that the mesh adapts to the image structure to give the best accuracy over the foreground pixels.

9: Image warping and spatial data mapping

4. Defining correspondence

The techniques described in this chapter take as input a series of 'tie-points', i.e. vertex pairs, one in each image, which define a set of points that should be brought into exact registration by the warp transform. These tie-points can be established by manual selection, automatic matching, or some combination. For the purposes of gene-expression mapping it is often the case that only part of the image can be properly matched and other parts have to be mapped onto other sections of the reference image. Furthermore, because the experimental section can suffer significant distortions, fully automatic matching is unlikely to be sufficiently reliable for the process of mapping data into the reconstruction. Therefore, it is necessary to be able to review the current set of tie-points and edit as necessary, including the deleting of putative pairs, and to be able to add new correspondences. For this, an interactive interface which shows the existing tie-point pairs as well as the result of warping using the current set is important. The prototype interface currently in use for this purpose includes three windows for displaying image data. The first two show the unwarped source and destination images and can accept pointer selection of tie-point vertices. The third window shows the warped source image overlaid on the destination image and is updated in real time as the tie-points are defined. Experience with a similar interface for registration (rotation and translation only) also indicates that the ability to mark points on a magnified window makes placement of the tie-points more accurate. The number of tie-point pairs that are required is application- and image-dependent. We have found that if the warp is smooth, in the sense that the displacement gradient is reasonably smooth, then relatively few (5–10) point correspondences are needed, whereas when there has been serious distortion many more are needed, some to delineate the structures that need to be moved and others to prevent the warp function from affecting tissues that do not need to be changed.

The tie-point program we have implemented to test the warping techniques has shown that certain features or properties of the interface are important for efficient user interaction (*Protocol 4*).

Protocol 4. Important tie-point interface properties and rules of thumb for tie-point selection

Properties
- Ability to delete and revise tie-point pairs
- A review display showing an overlay of the warped source and destination images
- Interactive update of the overlay display as tie-points are added
- Magnify options to allow more accurate placement of the tie-points

Rules of thumb
1. Select features of the images that constrain the tie-point in two directions such as corners or extrema of structures.

> **Protocol 4.** *Continued*
>
> 2. Use edge features after the primary alignment and warp has been achieved using unambiguous feature matches.
>
> 3. Only attempt 'reasonable' warp transformations. If the destination image structure has a very significant difference in spatial organization or topology then separately match the various parts as independent mappings.

5. Global transforms

Given that two images have been matched at only a finite number of points, a means of interpolating the displacements at those points to the rest of the image is required. The choice of function determines how well the original tie-point constraints are satisfied and the strength of the warp at regions distant from the tie-points. In this section we consider global functions, i.e. the interpolation functions extend to all space, in contrast to the local transforms discussed in the next section in which the functions only apply to a finite neighbourhood of the tie-points. The simplest functions have a fixed number of parameters which are determined by fitting to the displacements of the tie-points. If there are more tie-point constraints than parameters then the method described in Section 1.1 minimizes the sum of squared distances of the tie-point errors after transformation. The first three methods below are in this category: affine, polynomial, and conformal transformation.

More complex interpolation functions introduce basis functions or spline functions depending on the number of tie-point pairs. The parameters are determined by fitting each tie-point displacement exactly and by imposing constraints on the warp function, for example that it is continuous and smooth (continuous first derivative or gradient).

The first transform considered below is the affine transform which is a linear transform of the augmented coordinates (see below), and which in constrained mode is the simple 'registration transform' of translation and rotation but in the general case includes re-scaling and shear. When used for registration the transform does not warp the image in the usual sense of the word, but we include it here because a number of the techniques below are numerically more stable if the main translation, rotation, and sometimes scaling effects of the tie-point displacements are taken out by an affine pre-transformation.

To compare the results of the various global transforms considered here, we have used the same tie-point set, *Figure 3*, and calculated the warp transformation for each type. The results of these transformation can be compared by looking at the transformed meshes in *Figure 4* and the mapped and overlaid images in *Figure 5*.

9: Image warping and spatial data mapping

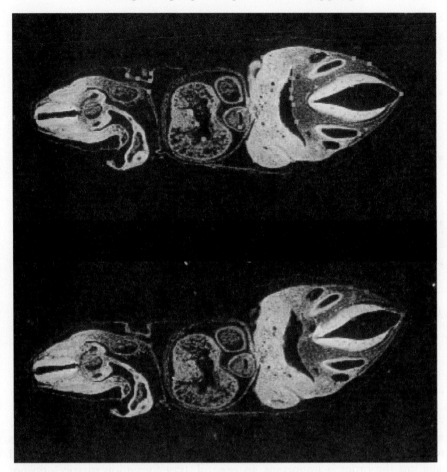

Figure 3. Two section images used as an example to compare the warp transformations. The red and blue dots show the tie-points set for this test.

5.1 Affine

An affine transform can be used to map one image to another in such a way that lines which are straight in the original image remain straight, and those that are parallel remain parallel. Given a point **p** in the original image and a point **q** in the transformed image, then the affine transform which maps **p** to **q** is **q** = **Ap** where **A** is the transformation matrix. In *homogeneous* coordinates (see Chapter 5) for 3D this is

$$\mathbf{q} = \begin{pmatrix} q_x \\ q_y \\ q_z \\ 1 \end{pmatrix} = \mathbf{Ap} = \begin{pmatrix} a_{11} & a_{12} & a_{13} & a_{14} \\ a_{21} & a_{22} & a_{23} & a_{24} \\ a_{31} & a_{32} & a_{33} & a_{34} \\ 0 & 0 & 0 & 1_{34} \end{pmatrix} \begin{pmatrix} p_x \\ p_y \\ p_z \\ 1 \end{pmatrix}. \qquad [10]$$

Often it is useful to constrain the affine transform to translation, rotation, scaling, or some combination of these such as registration (translation and rotation). In homogeneous coordinates these transform matrices are:

- 2D translation: $\mathbf{A_t} = \begin{pmatrix} 1 & 0 & t_x \\ 0 & 1 & t_y \\ 0 & 0 & 1 \end{pmatrix}$, where translation $\mathbf{t} = \begin{pmatrix} t_x \\ t_y \end{pmatrix}$.

- 3D translation: $\mathbf{A_t} = \begin{pmatrix} 1 & 0 & 0 & t_x \\ 0 & 1 & 0 & t_y \\ 0 & 0 & 1 & t_z \\ 0 & 0 & 0 & 1 \end{pmatrix}$, where translation $\mathbf{t} = \begin{pmatrix} t_x \\ t_y \\ t_z \end{pmatrix}$.

- 2D rotation: $\mathbf{A_r} = \begin{pmatrix} \cos(\theta) & -\sin(\theta) & 0 \\ \sin(\theta) & -\cos(\theta) & 0 \\ 0 & 0 & 0 \end{pmatrix}$ where θ is the rotation anticlockwise about the origin.

- 3D rotation: $\mathbf{A_r} = \mathbf{A_z A_y A_x}$, where:

$$\mathbf{A_z} = \begin{pmatrix} \cos(\theta) & -\sin(\theta) & 0 & 0 \\ \sin(\theta) & -\cos(\theta) & 0 & 0 \\ 0 & 0 & 1 & 0 \\ 0 & 0 & 0 & 1 \end{pmatrix}, \mathbf{A_y} = \begin{pmatrix} -\cos(\phi) & 0 & \sin(\phi) & 0 \\ -0 & 1 & 0 & 0 \\ -\sin(\phi) & 0 & \cos(\phi) & 0 \\ -0 & 1 & 0 & 0 \end{pmatrix},$$

and $\mathbf{A_x} = \begin{pmatrix} 1 & 0 & 0 & 0 \\ 0 & \cos(\varphi) & -\sin(\varphi) & 0 \\ 0 & \sin(\varphi) & \cos(\varphi) & 0 \\ 0 & 0 & 0 & 1 \end{pmatrix}.$

where θ, ϕ, and φ are the angles of rotation about the z, y, and x axes respectively. Note the transformation matrices can be multiplied in the usual way but the order is important. In general, transformation matrices do not commute, i.e. in general $\mathbf{AB} \neq \mathbf{BA}$.

- 2D scale: $\mathbf{A_s} = \begin{pmatrix} s_x & 0 & 0 \\ 0 & s_y & 0 \\ 0 & 0 & 1 \end{pmatrix}$, where s_x and s_y are the x and y scale factors respectively.

- 3D scale: $\mathbf{A_s} = \begin{pmatrix} s_x & 0 & 0 & 0 \\ 0 & s_y & 0 & 0 \\ 0 & 0 & s_z & 0 \\ 0 & 0 & 0 & 1 \end{pmatrix}$, with s_x, s_y, and s_z the x, y, z scale factors respectively.

To register a pair of images the affine transform which best maps one image onto the other has to be computed. This can be done either using image- or feature-based methods. Image-based methods include those based on finding the medial axis and centre of mass, maximum cross-correlation (12) and maximum mutual information (13, 14). Feature-based methods compute the

9: Image warping and spatial data mapping

'best fit' transform using tie-points identified in the two images. If the best fit is defined by minimizing the sum of squares of the tie-point distances, then solving for the transform coefficients reduces to a set of simultaneous equations which can be solved using standard linear algebra. The design equation for the coefficients of the new x-coordinate for the 3D affine transform with M tie-point pairs is

$$\begin{pmatrix} x_1 & y_1 & z_1 & 1 \\ x_2 & y_2 & z_2 & 1 \\ \vdots & \vdots & \vdots & \vdots \\ x_M & y_M & z_M & 1 \end{pmatrix} \begin{pmatrix} a_{11} \\ a_{12} \\ a_{13} \\ a_{14} \end{pmatrix} = \begin{pmatrix} u_1 \\ u_2 \\ \vdots \\ u_M \end{pmatrix} \qquad [11]$$

with similar equations for the y- and z-coordinate transforms.

If the required transform includes only translation and rotation (the Procrustes transform) then the minimization is of a non-linear set of equations.

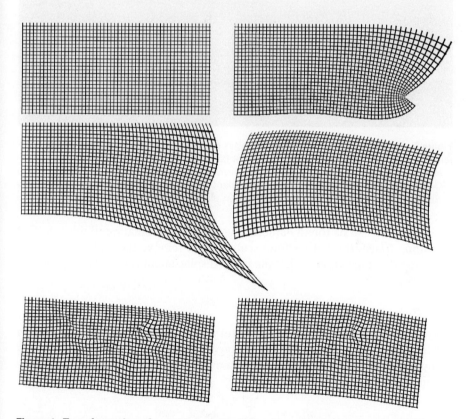

Figure 4. Transformation of a regular grid by the various global warp transforms. The top left shows the unwarped grid and traversing anticlockwise the grids correspond to: polynomial warp ($N = 3$); multiquadric; thin-plate spline; gaussian; and finally conformal transformation ($N = 6$). It is clear that the polynomial transformation is poorly constrained.

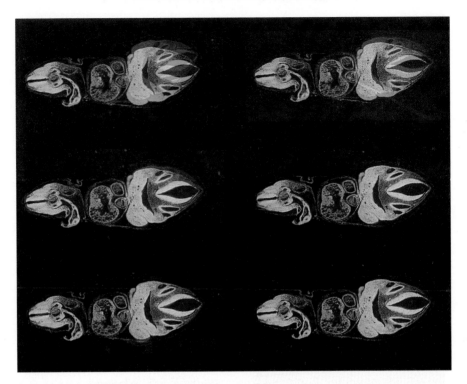

Figure 5. Transformation of the section images displaying the warped source image overlaid on the destination image using the colour comparator. The arrangement of the figures matches the order in *Figure 4*, anticlockwise from the top left: unwarped; polynomial warp ($N = 3$); multiquadric; thin-plate spline; gaussian; and finally conformal transformation ($N = 6$). In this case inspection shows that the thin-plate spline is the most robust.

The minimum can be found numerically; a fast approximation to this is to find the affine transform that includes scaling, then remove the scale component, adjusting the translation so that the mean displacement remains the same (see also Chapter 7).

5.2 Polynomial

Polynomial transforms have a long history of use in GISs and remote sensing applications where their chief advantage appears to have been their low computational cost. If the transform maps $I(x, y) \rightarrow I'(u, v)$ then the polynomial transform may be written as:

$$u(x, y) = \sum_{i=0}^{N_1-1} \sum_{j=0}^{N_2-1} a_{ij} x^i y^j \quad \text{and}$$
$$v(x, y) = \sum_{i=0}^{N_3-1} \sum_{j=0}^{N_4-1} b_{ij} x^i y^j.$$

[12]

9: *Image warping and spatial data mapping*

Usually, on the grounds of isotropy $N_1 = N_2 = N_3 = N_4 = N$. The polynomial coefficients a_{ij} are found by solving the design equation

$$\begin{pmatrix} 1 & x_1 & y_1 & \ldots & x_1^{N-1}y_1^{N-1} \\ 1 & x_2 & y_2 & \ldots & x_2^{N-1}y_2^{N-1} \\ 1 & \vdots & \vdots & \vdots & \vdots \\ \vdots & \vdots & \vdots & \vdots & \vdots \\ 1 & x_M & y_M & \ldots & x_M^{N-1}y_M^{N-1} \end{pmatrix} \begin{pmatrix} a_{00} \\ a_{10} \\ a_{01} \\ \vdots \\ a_{N-1 N-1} \end{pmatrix} = \begin{pmatrix} u_1 \\ u_2 \\ \vdots \\ \vdots \\ u_M \end{pmatrix} \quad [13]$$

where the values of x_i, y_i, and u_i are given by the tie-point correspondences between the images I and I'. The polynomial coefficients b_{ij} are found by solving the corresponding set of equations for the y destination coordinates. The examples shown in *Figures 4* and *5* were computed with polynomial order $N = 3$.

5.3 Conformal

A conformal transformation is a 2D transformation that preserves angles within the neighbourhood of any point in the sense that the angle of intersection of any two curves before and after transformation will remain the same in magnitude and direction. Conformal transformations can be used for re-mapping the entire image plane or for more local mapping of image values close to a given open or closed line. This has been used for straightening essentially 1D structures (e.g. chromosomes, see Chapter 3) or for re-mapping closed contours. The simplest conformal transformation to apply is in the form of a polynomial:

$$w = \sum_{k=0}^{N} a_k z^k, \quad [14]$$

where $z = x + iy$, $w = u + iv$, $i = \sqrt{-1}$, (x, y) are the coordinates in the original image, and (u, v) the coordinates in the warped image. The complex polynomial coefficients a_k are determined by substituting the given set of tie-points and solving the resultant set of simultaneous equations with standard linear algebra techniques. The design equation is:

$$\begin{pmatrix} \mathbf{A} & -\mathbf{B} \\ \mathbf{B} & -\mathbf{A} \end{pmatrix} \begin{pmatrix} \mathbf{\Lambda} \\ \mathbf{M} \end{pmatrix} = \begin{pmatrix} \mathrm{Re}(\mathbf{w}) \\ \mathrm{Im}(\mathbf{w}) \end{pmatrix}, \text{ where}$$

$$\mathbf{A} = \begin{pmatrix} 1 & \mathrm{Re}(z_1) & \mathrm{Re}(z_1^2) & \ldots & \mathrm{Re}(z_1^N) \\ 1 & \mathrm{Re}(z_2) & \mathrm{Re}(z_2^2) & \ldots & \mathrm{Re}(z_2^N) \\ \vdots & \vdots & \vdots & \vdots & \vdots \\ 1 & \mathrm{Re}(z_M) & \mathrm{Re}(z_M^2) & \ldots & \mathrm{Re}(z_M^N) \end{pmatrix}, \mathbf{B} = \begin{pmatrix} 0 & \mathrm{Im}(z_1) & \mathrm{Im}(z_1^2) & \ldots & \mathrm{Im}(z_1^N) \\ 0 & \mathrm{Im}(z_2) & \mathrm{Im}(z_2^2) & \ldots & \mathrm{Im}(z_2^N) \\ \vdots & \vdots & \vdots & \vdots & \vdots \\ 0 & \mathrm{Im}(z_M) & \mathrm{Im}(z_M^2) & \ldots & \mathrm{Im}(z_M^N) \end{pmatrix}, [15]$$

$$\mathbf{\Lambda} = \begin{pmatrix} \lambda_0 \\ \lambda_1 \\ \vdots \\ \lambda_N \end{pmatrix}, \mathbf{M} = \begin{pmatrix} \mu_0 \\ \mu_1 \\ \vdots \\ \mu_N \end{pmatrix}, \mathbf{W} = \begin{pmatrix} w_1 \\ w_2 \\ \vdots \\ w_M \end{pmatrix}, \text{ and } M \text{ is the number of tie-point pairs.}$$

If $M > N + 1$ it is not possible to have an exact match and the transformation will be approximate and the simplest option is to minimize the least squared distance error. If N is chosen large enough then the tie-points can be matched exactly, but as with the polynomial fitting the transformation may exhibit 'crazy' behaviour associated with singularities in the transformation. Note that if $N = 1$ the transformation is affine (preserves parallel lines). The example transform was calculated with $N = 6$.

There are two special cases worth mentioning which can be used to map lineal structures from one image to another, either as an open-ended or closed polyline. The essential difference from the more general mapping of tie-points is that some additional knowledge is used, namely that the points to be mapped are ordered, i.e. as a sequence of points along a curve with possibly the additional constraint that the first and last points are the same. This means that the curve on which the points lie will be preserved. In both cases the methods can be thought of as mapping onto a fixed line in an intermediate space; this line will either be a straight line (e.g. the abscissa) or a circle (e.g. the unit circle), in the case of an open or closed polyline respectively. In both cases the polynomial expansion can be considered as a parametric form of the input curve, but by using the orthogonal coordinate in the respective transformation the entire space can be traversed.

5.3.1 Mapping an open polyline

Given an ordered set of vertices (a polyline), $[z_j = (x_j, y_j): 1 \leq j \leq M]$, the proportionate distance of each point, s_j, along the line is calculated and used to generate the tie-points in the transformed space along the abscissa: $[w_j = (s_j/S, 0): 1 \leq j \leq M]$ where S is the total length. These can be plugged into the method above (but this time calculating the forward transform). However, because the destination tie-point coordinate is real, the simultaneous equations factorize into two independent calculations, one to match the x-coordinates and one to match the y-coordinates. By this method the original line has the parametric form:

$$x + iy = \sum_{k=0}^{N} a_k u^k \quad \text{where } 0 \leq u \leq 1 \qquad [16]$$

and a 'straightened' version of the image close to the line can be generated. Furthermore, for regions close to the polyline it does not matter if the curve passes back over itself (leaving aside the problem of disentangling the grey-values). However, for polylines that cross or have high curvature, the singularities associated with the reverse transformation mean that the transformation over the whole image may be ill-behaved. *Figure 6* shows the conformal coordinates (u, v) calculated for an arbitrary open polyline mapped back into (x, y) space. The order of the conformal polynomial was 6 and it is clear that when the original line has higher curvature, care must be taken not to stray too close to the singularities of the transformation.

9: Image warping and spatial data mapping

This technique can be used as a means of modelling the transformation of one open polyline onto another. If we have two polylines with fitted parametric forms:

$$z_1 = \sum_{k=0}^{N} a_{1k} u^k \quad \text{and} \quad z_2 = \sum_{k=0}^{N} a_{2k} u^k \qquad \text{where } 0 \leq u \leq 1, \qquad [17]$$

then introducing a new variable t allows us to define the surface swept out by the transformation from curve 1 to curve 2:

$$z(u,t) = \sum_{k=0}^{N} \left((1-t)a_{1k} + t a_{2k} u^k\right) \qquad \text{where } 0 \leq u \leq 1 \text{ and } 0 \leq t \leq 1 \qquad [18]$$

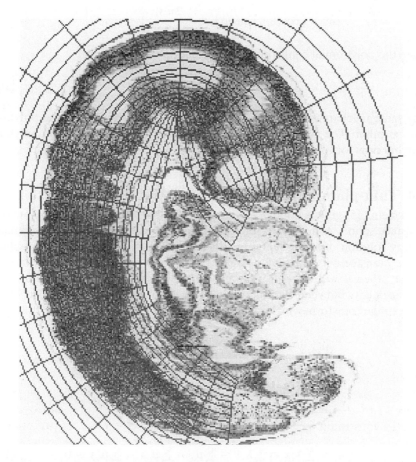

Figure 6. Conformal coordinate frame displayed on the original image. The line used to define the transformation was drawn roughly down the centre of the neural tissue. The section is a reconstructed view of the 9-day mouse embryo from the Edinburgh Mouse Atlas.

Alternative functions, other than a simple linear transform, to describe the path from line 1 to line 2 are of course possible, for example $(\cos(\theta)a_{1k} + \sin(\theta)a_{2k})$ which would define a cyclic change between line 1 and line 2.

5.3.2 Mapping a closed polyline

If the polyline is closed then mapping the line onto the abscissa will result in a singularity in the reverse transform somewhere in the region inside the polyline. To avoid this and allow mapping of all the pixels within the polyline, it is appropriate to match the polyline onto the unit circle $w = re^{i\theta}$ with $r = 1$. The match points on the unit circle for the set of vertices on the closed polyline are

$$\left[w_j = (\cos(\theta_j), \sin(\theta_j)), \theta_j = \frac{2\pi s_j}{S} : 1 \leq j \leq M\right]. \quad [19]$$

Then the polynomial conformal transformation takes the form:

$$z = (x + iy) = \sum_{k=0}^{N} a_k r^k \, e^{ik\theta}. \quad [20]$$

The same mechanism as above can be used to transform one closed polyline onto another. The conformal coordinate grid is similar to the open polyline case but now forms a radial structure.

5.4 Thin-plate spline

Bookstein (15) gives a good description of the thin-plate spline, also known as a surface spline. A thin-plate spline is so called because it models a minimum energy solution of a thin metal plate subjected to point loads. If a thin-plate spline transform is calculated from a set of tie-points between a pair of images, then it will map all tie-points of the first image to the second, with displacements between the tie-points being smoothly interpolated. The thin-plate spline transform which maps image $I(x, y) \to I'(u, v)$ may be written as:

$$u(x, y) = a_0 + a_1 x + a_2 y + \tfrac{1}{2}\sum_{i=1}^{N}\lambda_i r_i^2 \log(r_i^2)$$
$$v(x, y) = b_0 + b_1 x + b_2 y + \tfrac{1}{2}\sum_{i=1}^{N}\mu_i r_i^2 \log(r_i^2) \quad [21]$$

with the constraints

$$\sum_i \lambda_i = \sum_i \lambda_i x = \sum_i \lambda_i y = \sum_i \mu_i = \sum_i \mu_i x = \sum_i \mu_i y = 0, \quad [22]$$

where

$$r_i^2 = (x_i - x)^2 + (y_i - y)^2 \quad [23]$$

9: Image warping and spatial data mapping

and (x_i, y_i) is the position of the ith tie-point in the first image. The design equation for the thin-plate spline transform is:

$$\begin{pmatrix} 0 & 0 & 0 & 1 & 1 & \cdots & 1 \\ 0 & 0 & 0 & x_1 & x_2 & \cdots & x_N \\ 0 & 0 & 0 & y_1 & y_2 & \cdots & y_N \\ 1 & x_1 & y_1 & 0 & r_{12}^2 \log r_{12}^2 & \cdots & r_{1N-1}^2 \log r_{1N}^2 \\ 1 & x_2 & y_2 & r_{21}^2 \log r_{21}^2 & 0 & \cdots & r_{2N-1}^2 \log r_{2N}^2 \\ \vdots & \vdots & \vdots & \vdots & \vdots & & \vdots \\ 1 & x_N & y_N & r_{N1}^2 \log r_{N1}^2 & r_{N2}^2 \log r_{N2}^2 & \cdots & 0 \end{pmatrix} \begin{pmatrix} a_0 \\ a_1 \\ a_2 \\ \lambda_1/2 \\ \lambda_2/2 \\ \vdots \\ \lambda_n/2 \end{pmatrix} = \begin{pmatrix} 0 \\ 0 \\ 0 \\ z_1 \\ z_2 \\ \vdots \\ z_N \end{pmatrix} \quad [24]$$

where

$$r_{ij}^2 = r_{ji}^2 = (x_i - x_j)^2 + (y_i - y_j)^2 \quad [25]$$

for some pair of tie-points i,j and $z_i = u_i - x_i$ is the x-component of the tie-point displacement, and $z_i = v_i - y_i$ in the corresponding equation for b_i, μ_i.

Solving *Equation 24* can in some cases be difficult because of numerical error if the matrix is poorly conditioned. Following Barrodale *et al.* (11) and Franke (16), the condition number can be significantly reduced by re-scaling the tie-point coordinate values with respect to the maximal spatial range of the tie-point values. Therefore, a more stable numerical solution can be found using:

$$\hat{x}_i = \frac{1}{C}\left(x_i - \min_i x_i\right), \quad \hat{y}_i = \frac{1}{C}\left(y_i - \min_i y_i\right), \quad [26]$$

where

$$C = \max\left((\max_i x_i - \min_i x_i), (\max_i y_i - \min_i y_i)\right), \quad [27]$$

i.e. C is the maximal spatial range of the tie-points. Barrodale *et al.* (11) report that this re-scaling allows the determination of thin-plate spline transforms using up to about 2000 tie-points.

Equation 24 is solved with standard linear algebra algorithms using the scaled coordinates. The unscaled coefficients are computed as follows:

$$a_0 = \hat{a}_0 - \frac{\hat{a}_1}{C} \min_i x_i - \frac{\hat{a}_2}{C} \min_i y_i - 2 \log(C) \sum_{i=1}^{N} \hat{\lambda}_i (\hat{x}_i^2 + \hat{y}_i^2),$$

$$a_1 = \frac{\hat{a}_1}{C}, a_2 = \frac{\hat{a}_2}{C}, \text{ and } \lambda = \frac{\hat{\lambda}_i}{C} \text{ for } i = 1, 2 ..., N, \quad [28]$$

The coefficients b_i and μ_i are calculated in similar fashion.

For a thin-plate spline transform computed from N tie-points $2N$ log terms need to be computed for each transformed pixel, making the thin-plate spline computationally expensive. Barrodale *et al.* (11) make use of an approximation algorithm due to Powell (10) which is reported to give approximately a two orders of magnitude improvement in execution time. We have found similar improvements in efficiency using the triangulation, piecewise linear method described in Section 3.

5.5 Multiquadric

The multiquadric, like the thin-plate spline, is a radial basis function. A multiquadric mapping $I(x, y) \to I'(u, v)$ may be written

$$u(x, y) = a_0 + a_1 x + a_2 y + \sum_{i=1}^{N} \lambda_i \sqrt{r_i^2 + R^2}$$
$$v(x, y) = b_0 + b_1 x + b_2 y + \sum_{i=1}^{N} \mu_i \sqrt{r_i^2 + R^2}$$
[29]

where r_i is the distance from the ith tie-point to (x, y) and R is an arbitrary constant. There are many recommendations for choosing values of R (17); however, we have observed that a value given by

$$R = 10 \max \left((\max_i x_i - \min_i x_i), (\max_i y_i - \min_i y_i) \right),$$
[27]

produces good results.

The design equation for the multiquadric transform is similar to that of the thin-plate spline:

$$\begin{pmatrix} 0 & 0 & 0 & 1 & 1 & \dots & 1 \\ 0 & 0 & 0 & x_1 & x_2 & \dots & x_N \\ 0 & 0 & 0 & y_1 & y_2 & \dots & y_N \\ 1 & x_1 & y_1 & \sqrt{r_{11}^2+R^2} & \sqrt{r_{12}^2+R^2} & \dots & \sqrt{r_{1N}^2+R^2} \\ 1 & x_2 & y_2 & \sqrt{r_{21}^2+R^2} & \sqrt{r_{22}^2+R^2} & \dots & \sqrt{r_{2N}^2+R^2} \\ \vdots & \vdots & \vdots & \vdots & \vdots & \vdots & \vdots \\ 1 & x_N & y_N & \sqrt{r_{N1}^2+R^2} & \sqrt{r_{N2}^2+R^2} & \dots & \sqrt{r_{NN}^2+R^2} \end{pmatrix} \begin{pmatrix} a_0 \\ a_1 \\ a_2 \\ \lambda_1 \\ \lambda_2 \\ \vdots \\ \lambda_N \end{pmatrix} = \begin{pmatrix} 0 \\ 0 \\ 0 \\ z_1 \\ z_2 \\ \vdots \\ z_N \end{pmatrix}$$
[31]

where r_{ij} and z_i are defined by *Equation 25*. The same re-scaling used for the thin-plate spline may also be useful for solving for the multiquadric parameters and the unscaled coefficients can be recovered using:

$$a_0 = \hat{a}_0 - \frac{\hat{a}_1}{C} \min_i x_i - \frac{\hat{a}_2}{C} \min_i y_i,$$
$$a_1 = \frac{\hat{a}_1}{C}, a_2 = \frac{\hat{a}_2}{C}, \text{ and } \lambda = \frac{\hat{\lambda}_i}{C} \text{ for } i = 1, 2, \dots, N,$$
[32]

5.6 Gaussian

A Gaussian radial basis function may be defined similarly to the thin-plate spline and multiquadric:

$$u(x, y) = a_0 + a_1 x + a_2 y + \sum_{i=1}^{N} \lambda_i \exp\left(\frac{r_i^2}{R^2}\right)$$
$$v(x, y) = b_0 + b_1 x + b_2 y + \sum_{i=1}^{N} \mu_i \exp\left(\frac{r_i^2}{R^2}\right)$$
[33]

where R is a constant.

Calculation of the transform coefficients also follows similarly; however, we have not managed to achieve useful results using the Gaussian radial basis function.

Protocol 5. Calculating the transform parameters using the design equation

Algorithms
- Method to define the tie-points
- Method to calculate the basis function
- Linear algebra—singular value decomposition and back-substitution (5)

Method
1. For each tie-point pair, use the source vertex to calculate the row of basis function values in the matrix and use the destination x-value for the RHS constraint vector.
2. Use singular value decomposition to convert the design matrix to the appropriate form. Set values close to zero (e.g. magnitude less than 10^{-6}) to be exactly zero.
3. Calculate the coefficients using back-substitution.
4. Repeat for the y-transformation coefficients using the same computed matrix.

---

6. Local transforms

Local transforms are those for which the effect of warping at a given point is strictly constrained to a neighbourhood of that point. This allows a much more complex re-mapping of the image but will typically require much more detailed input. The methods we illustrate are based on a predefined mesh which in our case is triangular, but in others (18) rectangular. In this section we describe two methods. The first is to use the tie-points to displace a given node in the mesh with the new image calculated by interpolating within the mesh elements that contain that node. The second method is to impose a physical model on the mesh and establish the warp transform by solving a set of equations describing the 'physical' behaviour. In our case we describe the use of the finite element method (FEM) to find the warp of the image assuming it behaves like a hyper-elastic membrane.

6.1 Simple mesh transformation

This is the simplest option for warping. The image is divided up into a mesh of triangles or rectangles and the nodes of the mesh moved as required. The movement of each node will affect only a subset of the mesh elements and the warped image is calculated by interpolation over that set. In the simplest case the interpolation is bilinear over the affected triangles. In the case of *xmorph* (18) the interpolation uses a 2D spline. In either case, since the movement of a node does not move any others, the range of movement must be restricted to the maximum inscribed circle centred within the polygon defined by the edges of the mesh elements that the include the given node, but not those edges passing through the node (see *Figure 2*). This is to avoid mesh elements (triangles or quadrilaterals) being reduced to zero area or 'flipped' over which would be the equivalent of a fold in the image.

6.2 Elastic plate warping

For the purposes of 3D reconstruction, Guest and Baldock (19) have developed a technique based on the idea that the image to be warped can be modelled as a thin elastic plate. The matching between adjacent images in the stack result in 'forces' being applied to the elastic plate. The plate is then allowed to deform until the externally applied forces (from the matching) are balanced by the internal forces arising from the deformation (strain) of the elastic material. This equilibrium point is found using the techniques of the FEM. Because the elastic properties of the image resist deformation, if used for the purposes of tie-point matching the tie-points could not be brought into exact registration. For this purpose, therefore, the method can be modified either by letting the external forces become very large and thereby overwhelm any internal forces, or by progressively reducing the coefficients of elasticity of the material. This method has the advantage that good results can be achieved with fewer tie-points, but with the cost that the computation time is currently too high (minutes per image) to be useful within an interactive system.

7. Mapping gene-expression data

These warping techniques have been implemented to enable spatial mapping of gene-expression data from sections of an experimental embryo onto a standard reference embryo. In the case of mouse data the standard reference is the Edinburgh Mouse Atlas (4) which has a set of 3D voxel models from which the corresponding section can be obtained and then used as the destination image for the warping. The warp transform is defined interactively using tie-points and then the gene-expression data mapped onto the standard. The data are typically extracted from the original image by simple thresholding (see Chapter 2) of the signal image. To extract the required section view from the 3D voxel image we use the program MAPaint which was developed for the task of defining the anatomical domains. This program allows section views at arbitrary orientation and positions through the stack and therefore it is possible to select the section which best matches the

9: Image warping and spatial data mapping

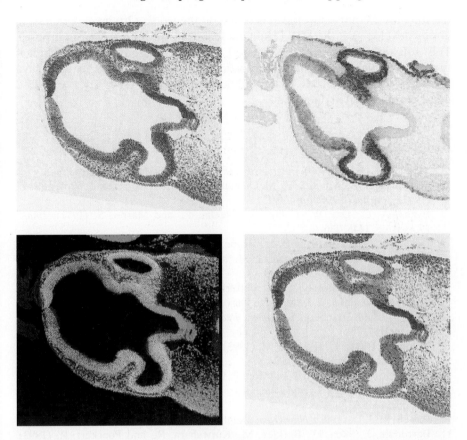

Figure 7. Mapping gene-expression data from a section from the experimental embryo (top right) to the corresponding 'Atlas' section (top left). The transformed source image is overlaid on the atlas image (bottom left) and the transformation used to map the extracted (simple threshold) gene-expression data onto the atlas section (bottom right).

experimental data. The matching section located in the atlas is the destination image for the source data in the experimental image. This is mapped using a tie-point interface which interactively updates the warp for each tie-point added, so that at each step the quality of the current mapping can be established. When sufficient tie-points have been placed, the experimental data are warped and used to define the gene-expression domain within the atlas voxel model. A result of this process is illustrated in *Figure 7*.

8. Summary

In this chapter we have introduced a number of methods for warping 2D images as well as the associated techniques required for a usable system. All the software we have developed is based on the *woolz* (20) image processing library developed at the MRC Human Genetics Unit for a range of purposes. This software library and associated X11/Motif-based GUI software is currently available for bone fide academic research; the library and Java interfaces will be available in the public domain (with GNU style 'copyleft' usage permissions) in due course.

References

1. Thompson, D. W. (1969). *On growth and form*. Cambridge University Press.
2. Scarborough, J., Aiton, J. F., McLachlan, J. C., Smart, S. D., and Whiten, S. C. (1997). *Journal of Anatomy*, **191**, 117.
3. Evans, A. C., Dai, W., Collins, L., Neelin, P., and Marret, S. (1991). *Image Processing*, **1445**, 236.
4. Davidson, D. R., Bard, J. B. L., Brune, R., Burger, A., Dubreuil, C., Hill, W., Kaufman, M. H., Quinn, J., Stark, M., and Baldock, R. (1997). *Seminars in Cell and Development Biology*, **8**, 509.
5. Press, W. H., Teukolsky, S. A., Vetterling, W. T., and Flannery, B. P. (1992). *Numerical recipes in C*. Cambridge University Press.
6. Floyd, R. and Steinberg, L. (1976). *Proceedings of the Society for Information Display*, **117**, 75.
7. Neider, J., Davis, T,. and Woo, M. (1993). *OpenGL programming guide*. Addison-Wesley, Menlo Park (CA), USA.
8. Zimmermann, M. H. and Tomlinson, P. B. (1966). *Science*, **152**, 72.
9. Wolberg, G. (1992). *Digital image warping*. IEEE Computer Society Press, Los Alamitos, USA.
10. Powell, M. J. D. (1992). *Numerical Methods of Approximation Theory*, **9**, 221.
11. Barrodale, J., Skea, D., Berkley, M., Kurwahara, R., and Poeckert, R. (1993). *Pattern Recognition*, **26**, 375.
12. Hibbard, L. S., Grothe, R. A., and Arnicar-Sulze, T. L. (1993). *SPIE*, **1905**, 946.
13. Viola, P. and Wells, W. M. (1997). *Int. J. of Computer Vision*, **24**, 137.
14. Studholme, C., Hill, D. L. G., and Hawkes, D. J. (1994). *Proc. 5th BMVC Conference*, pp. 235. BMVA Press, UK.
15. Bookstein, F. L. (1989). *IEEE Trans. on Pattern Analysis and Machine Intelligence*, **11**, 567.
16. Franke, R. (1981). *Smooth interpolation of scattered data by local thin plate splines*. Naval Postgraduate School, Monterey, USA.
17. Carlson, R. E. (1991). *Computers and Mathematics with Applications*, **21**, 29.
18. http://www.colorado-research.com/~gourlay/software/Graphics/Xmorph
19. Guest, E. and Baldock, R. (1995). *Bioimaging*, **3**, 154.
20. Piper, J. and Rutovitz, D. (1985). *Pattern Recognition Letters*, **3**, 119.

A1
List of suppliers

AIR, http://bishopw.loni.ucla.edu/AIR3
Amerix Applied Imaging Inc., http://www.aai.com
Amersham
Amersham International plc., Lincoln Place, Green End, Aylesbury, Buckinghamshire HP20 2TP, UK.
Amersham Corporation, 2636 South Clearbrook Drive, Arlington Heights, IL 60005, USA.
Analytical Imaging Station, Imaging Research, http://imaging.brocku.ca
Anderman
Anderman and Co. Ltd., 145 London Road, Kingston-Upon-Thames, Surrey KT17 7NH, UK.
Andover Corporation, Salem, NH 03079, USA.
Beckman Instruments
Beckman Instruments UK Ltd., Oakley Court, Kingsmead Business Park, London Road, High Wycombe, Bucks HP11 1J4, UK.
Beckman Instruments Inc., PO Box 3100, 2500 Harbor Boulevard, Fullerton, CA 92634, USA.
Becton Dickinson
Becton Dickinson and Co., Between Towns Road, Cowley, Oxford OX4 3LY, UK.
Becton Dickinson and Co., 2 Bridgewater Lane, Lincoln Park, NJ 07035, USA.
Bellcore Customer Service, 8 Corporate Place, Piscataway, NJ 08554, USA, www.research.att.com/~andreas/xgobi
Bio
Bio 101 Inc., c/o Statech Scientific Ltd, 61–63 Dudley Street, Luton, Bedfordshire LU2 0HP, UK.
Bio 101 Inc., PO Box 2284, La Jolla, CA 92038–2284, USA.
Bio-Rad Laboratories
Bio-Rad Laboratories Ltd., Bio-Rad House, Maylands Avenue, Hemel Hempstead HP2 7TD, UK.
Bio-Rad Laboratories, Division Headquarters, 3300 Regatta Boulevard, Richmond, CA 94804, USA.

List of suppliers

Boehringer Mannheim
Boehringer Mannheim UK (Diagnostics and Biochemicals) Ltd, Bell Lane, Lewes, East Sussex BN17 1LG, UK.
Boehringer Mannheim Corporation, Biochemical Products, 9115 Hague Road, P.O. Box 504 Indianapolis, IN 46250–0414, USA.
Boehringer Mannheim Biochemica, GmbH, Sandhofer Str. 116, Postfach 310120 D-6800 Ma 31, Germany.
British Drug Houses (BDH) Ltd, Poole, Dorset, UK.
Carl Zeiss, 73446 Oberkochen, Germany.
Computational Imaging Science Group, http://carmen.umds.ac.uk/cisg/software
CVIP Tools, University of Southern Illinois, USA, http://www.ee.siue.edu/CVIPtools
Difco Laboratories
Difco Laboratories Ltd., P.O. Box 14B, Central Avenue, West Molesey, Surrey KT8 2SE, UK.
Difco Laboratories, P.O. Box 331058, Detroit, MI 48232–7058, USA.
Du Pont
Dupont (UK) Ltd., Industrial Products Division, Wedgwood Way, Stevenage, Herts, SG1 4Q, UK.
Du Pont Co. (Biotechnology Systems Division), P.O. Box 80024, Wilmington, DE 19880–002, USA.
Edmund Scientific, http://www.edsci.com
European Collection of Animal Cell Culture, Division of Biologics, PHLS Centre for Applied Microbiology and Research, Porton Down, Salisbury, Wilts SP4 0JG, UK.
Falcon (Falcon is a registered trademark of Becton Dickinson and Co.).
Fisher Scientific Co., 711 Forbest Avenue, Pittsburgh, PA 15219–4785, USA.
Flow Laboratories, Woodcock Hill, Harefield Road, Rickmansworth, Herts. WD3 1PQ, UK.
Fluka
Fluka-Chemie AG, CH-9470, Buchs, Switzerland.
Fluka Chemicals Ltd., The Old Brickyard, New Road, Gillingham, Dorset SP8 4JL, UK.
Genius, http://www.genius-kye.com
Gibco BRL
Gibco BRL (Life Technologies Ltd.), Trident House, Renfrew Road, Paisley PA3 4EF, UK.
Gibco BRL (Life Technologies Inc.), 3175 Staler Road, Grand Island, NY 14072–0068, USA.
Arnold R. Horwell, 73 Maygrove Road, West Hampstead, London NW6 2BP, UK.
Hayden Image Processing Group, http://www.perceptive.com
Hybaid
Hybaid Ltd., 111–113 Waldegrave Road, Teddington, Middlesex TW11 8LL, UK.

List of suppliers

Hybaid, National Labnet Corporation, P.O. Box 841, Woodbridge, NJ. 07095, USA.
HyClone Laboratories 1725 South HyClone Road, Logan, UT 84321, USA.
ImageTool, http://ddsdx.uthscsa.edu/dig/itdesc.html
International Biotechnologies Inc., 25 Science Park, New Haven, Connecticut 06535, USA.
Invitrogen Corporation
Invitrogen Corporation 3985 B Sorrenton Valley Building, San Diego, CA. 92121, USA.
Invitrogen Corporation c/o British Biotechnology Products Ltd., 4–10 The Quadrant, Barton Lane, Abingdon, OX14 3YS, UK.
Khoral Research Inc., 6200 Uptown Bvd. N.E., Albuquerque, NH. http://www.khoral.com
Kodak: Eastman Fine Chemicals 343 State Street, Rochester, NY, USA.
Leica Microsystems (UK) Ltd, Davy Avenue, Knowlhill, Milton Keynes MK5 8LB, UK.
Life Technologies Inc., 8451 Helgerman Court, Gaithersburg, MN 20877, USA.
LView Pro, http://www.lview.com
The Mathworks Inc, 24 Prime Park Way, Natick, MA 01760-1500, USA, www.mathworks.com
Media Cybernetics, http://www.mediacy.com
Merck
Merck Industries Inc., 5 Skyline Drive, Nawthorne, NY 10532, USA.
Merck, Frankfurter Strasse, 250, Postfach 4119, D-64293, Germany.
Millipore
Millipore (UK) Ltd., The Boulevard, Blackmoor Lane, Watford, Herts WD1 8YW, UK.
Millipore Corp./Biosearch, P.O. Box 255, 80 Ashby Road, Bedford, MA 01730, USA.
Molecular Probes, Inc., PO Box 22010, Eugene, OR 97402-0469, USA.
The Neural Computing Research Group, Aston University, United Kingdom, www.ncrg.aston.ac.uk/netlab/over.html
NeuralWare Inc., 202 Park West Drive, Pittsburgh, PA 15275, USA, www.neuralware.com
New England Biolabs (NBL)
New England Biolabs (NBL), 32 Tozer Road, Beverley, MA 01915–5510, USA.
New England Biolabs (NBL), c/o CP Labs Ltd., P.O. Box 22, Bishops Stortford, Herts CM23 3DH, UK.
NIH Image, http://rsb.info.nih.gov/nih-image
Nikon Corporation, 9-16, Ohi 3-chome, Shinagawa-ku, Tokyo 140, Japan.
Noesis Vision Ltd, http://www.cam.org/~noesis
Olympus Optical Co., **Ltd**, PO Box 7004, Shinjuku Monolith Building, 2-3-1 Nishi-Shinjuku, Shinjuku-ku Tokyo, 163-0, Japan.

List of suppliers

Perkin-Elmer
Perkin-Elmer Ltd., Maxwell Road, Beaconsfield, Bucks. HP9 1QA, UK.
Perkin-Elmer Ltd., Post Office Lane, Beaconsfield, Bucks, HP9 1QA, UK.
Perkin-Elmer-Cetus (The Perkin-Elmer Corporation), 761 Main Avenue, Norwalk, CT 0689, USA.
Pharmacia Biotech Europe Procordia EuroCentre, Rue de la Fuse-e 62, B-1130 Brussels, Belgium.
Pharmacia Biosystems
Pharmacia Biosystems Ltd. (Biotechnology Division), Davy Avenue, Knowlhill, Milton Keynes MK5 8PH, UK.
Pharmacia LKB Biotechnology AB, Björngatan 30, S-75182 Uppsala, Sweden.
Photometrics Ltd., Tucson, AZ, USA.
Promega
Promega Ltd., Delta House, Enterprise Road, Chilworth Research Centre, Southampton, UK.
Promega Corporation, 2800 Woods Hollow Road, Madison, WI 53711–5399, USA.
Qiagen
Qiagen Inc., c/o Hybaid, 111–113 Waldegrave Road, Teddington, Middlesex, TW11 8LL, UK.
Qiagen Inc., 9259 Eton Avenue, Chatsworth, CA 91311, USA.
Registration SPECT/PET & MR, http://www.fil.ion.ucl.ac.uk/spm
The SAS Institute Inc., SAS Campus Drive, Cary, NC 27513–2414, USA, www.sas.com
Schleicher and Schuell
Schleicher and Schuell Inc., Keene, NH 03431A, USA.
Schleicher and Schuell Inc., D-3354 Dassel, Germany. Schleicher and Schuell Inc., c/o Andermann and Company Ltd.
SCIL-Image, TNO, The Netherlands, http://www.tpd.tno.nl
Scion Inc., USA, http://www.scioncorp.com
Shandon Scientific Ltd., Chadwick Road, Astmoor, Runcorn, Cheshire WA7 1PR, UK.
Sigma Chemical Company
Sigma Chemical Company (UK), Fancy Road, Poole, Dorset BH17 7NH, UK.
Sigma Chemical Company, 3050 Spruce Street, P.O. Box 14508, St. Louis, MO 63178–9916.
Soft Imaging Systems GmbH, Hammer Str. 89, D-48153 Muenster, Germany.
Sony Electronics Inc., Park Ridge, NJ, USA.
Sorvall DuPont Company, Biotechnology Division, P.O. Box 80022, Wilmington, DE 19880–0022, USA.
Spherotech, Inc., 1840 Industrial Dr. Suite 270, Libertyville, IL 60048-9817, USA.
SPSS Inc., 2333 S. Wacker Drive, Chicago, IL 60606-6307, USA, www.spss.com

List of suppliers

Stratagene
Stratagene Ltd., Unit 140, Cambridge Innovation Centre, Milton Road, Cambridge CB4 4FG, UK.
Strategene Inc., 11011 North Torrey Pines Road, La Jolla, CA 92037, USA.
Summagraphics, http://www.summagraphics.com
Sun Microsystems Ltd., http://www.sun.com
TIM for Windows, http://www.ph.tn.tudelft.nl/software.html
United States Biochemical, P.O. Box 22400, Cleveland, OH 44122, USA.
Visual Automation Ltd, http://www.isbe.man.ac.uk/VAL
Wacom, http://www.wacom.com
Wellcome Reagents, Langley Court, Beckenham, Kent BR3 3BS, UK.
WiT, Logical Vision Ltd, http://www.logicalvision.com
Woolz IP Software, http://www.genex.hgu.mrc.ac.uk

Index

3D *see* three-dimensional
4-connected *see* connectivity
8-connected *see* connectivity
active contour 85–87
  greedy algorithm 85
  energy function 86
  models 225
affine *see* transformations
Airy disc 3
Alice 106
alignment 242–244
  Procrustes 242
  of sections 166
  tangent space 243
  see also iterative closest point; registration
Analytical Imaging Station 105
AND *see* binary image processing
Aphelion 106
aspect ratio 125
astigmatism 12
atlas 198, 212, 261
  see also Edinburgh Mouse atlas
augmented coordinates 274
  see also homogenous coordinates
autofluorescence 15
autofocus 31, 104
automation *see* computer controlled microscope
  automatic karotyping 69
  quantitative microscopy 1

banding *see* chromosome
barrel distortion 12
basis functions 263
  radial, network 134–135
  radial, interpolation 284
  sinusoidal (Fourier) 100
Bayes'
  risk 115
  rule 97, 114
Bayesian classification 96–98
binary image 87
  overlay 264–7
binary image processing 87
  cross-correlation 178
  hit-or-miss transform 87
  logical AND 89, 266
  logical OR 89, 266
  overlap 179
  see also mathematical morphology
binomial distribution 98
blink comparator 264, 268–269

bootstrap method 141, 142
  see also classifier validation
boundary refinement 72, 85–94
bounding box 263
Buffon's needle 253

camera 1, 19–28
  choosing 27
  calibration 23
  see also charge coupled device
Canny edge detector *see* edge
CCD *see* charge coupled device
centre of gravity 203
centromere 70, 94 113
centromeric index 94, 125
CGH *see* comparative genomic hybridisation
chamfer distance 205
charge coupled device (CCD) 19–24
  binning 23, 32
  cooling 26, 30
  fill factor 21
  gain 22
  gamma 22
  linearity 22
  quantum efficiency 22
  sensitivity 21
  spectral sensitivity 21
  see also camera
chromatic aberration 11
chromosome
  analysis 69
  banding 94
  classification *see* classification
  features
    banding pattern 94–96
    morphological 94
  measurement 93
  painting 102
  Q-banding 94
  straightening 99
  X and Y 113
chromosome domains
  location 257
  size of 259
circularity 125
city-block *see* distance
classification 111
  of chromosomes 96, 112–113, 137–140
  colour 103
  features 94–96
  misclassification cost 115, 117
  probabilistic framework for 114
  space 117

Index

classifier
  Bayesian 96–98
  maximum likelihood 114
  box 115
  linear 117–121
  quadratic 117–121
  non-parametric 121
  parametric 121
  nearest neighbour 122
  k-Nearest neighbour 123, 124
  training 98, 112
    set size 143
  validation 141
closing *see* mathematical morphology
clustering 129–131
colour comparator 267–268
coma 12
compactness 125
comparative genomic hybridisation (CGH) 70, 102
computed tomography (CT) 38, 39, 154, 261
computer assisted reconstruction (CAR) 153
computer controlled microscope stage 1, 30
confocal microscope (CLSM) 13, 155
confusion matrix 142
congruencing 168, 180
conjugate planes 8
connectivity 72, 227
  4-connected 72
  8-connected 72
contour 159
  annotation 165
  acquisition 164
  software 165
contrast manipulation 37, 39–46
convex hull 92
convolution 77
  Gaussian 78
  kernel 77
  smoothing 78
coordinate transformation 261
correlation 224
  see also cross-correlation
correspondence
  tie-points 273
cost-function 232, 233
covariance matrix 228, 235
cross validation 141
cross-correlation 179, 185, 276
  normalised 208
CT *see* computed tomography
curvature
  3D 203
  analysis 91, 93
  of field 12
curve fitting
  conformal polyline 280
  for edge linking 82

iterative endpoint fitting algorithm 82
  piecewise linear 83
  see also Hough transform
cytogenetics 69
  see also karotyping
cytophotometry 2

data fusion *see* image fusion
decision space 117
decision surface 115, 117
  linear 119
  piecewise linear 122
  quadratic 119
deformable models 225
density profile 113
depth of field 10
design equation 263, 282–285
  for multiquadric 284
  for thin-plate spline 283
developmental biology 154
diaphragm 8, 17
diascopic 190, 191
digitizing tablet 161
  LCD tablet 164
dilation *see* mathematical morphology
discriminant function 119
distance
  city-block 123
  Euclidean 123
  Mahalanobis 98, 118, 123
distortion *see* optical distortion
dithering 264–6
DNA loci 256

edge 77–84, 233
  Canny edge detector 51, 81
  detection 49, 51, 77–81
  Kirsch operator 80
  Laplacian 77
  Laplacian of Gaussian 78
  linking 78, 81–84
  multiresolution edge detection 81
  Roberts operator 80
  Sobel operator 49, 51, 80
  Prewitt operator 80
  scale 51
Edinburgh Mouse Atlas 286–287
eigen
  analysis 128
  vector 128, 228
  see also principal components
elastic plate warping 286
energy function
  for active contour 86

296

Index

episcopic image 190, 191
erosion *see* mathematical morphology
Euclidean *see* distance
expectation-maximisation (EM) algorithm 121

face image 224
factor analysis 128
false positive, negative 144–145
feature 1
   clustering 129–131
   combining 128–9
   correlation of 125, 128
   non-commensurate 123
   selection 113, 124–128
   space 117
   vector 114
fiducial 168–73
   landmark frames 171
   markers 199
   point 168–9
   region 168–71
   see also tie-points
filter
   image filtering 37, 46–53
      edge detection 49
      fourier representation 49–50
      Gaussian 47
      gradient 76
      high-pass 100
      linear 46–50
      median 50
      morphological 46,
      non-linear 47, 50–53
      sharpening 48
      smoothing 47
      spatially adaptive 50
      unsharp mask 49
      see also image transforms
   optical 17–18
finite element method 285
FISH 70, 102
   M-FISH 70, 103
fit-function *see* cost-function
flare 6
Floyd-Steinberg *see* dithering
fluorescence 2, 6
   in situ hybridisation 70, 102
      multicolour 70
      multiplex 103
   microscopy 2, 13
   see also autofluorescence
Fourier
   coefficients 94
   descriptors 224
   fast transform 186
   transform 100
      for filtering 49–50

frame grabber 24, 162
frequency domain 49
FROC *see* receiver operating characteristic
functional transformation 40
fungal mycelia 37–38
fusion *see* image fusion

Gaussian function
   in Canny edge detector 51
   as chromosome classification feature 95
   convolution 78
   Laplacian of 78
   as normal distribution 98
   separable 47
   as smoothing filter 47
gene-expression 154, 286
genetic algorithm (GA) 127, 232
geometric distortion 12
geometric transformation 173
   linear 173
   rigid 173
   spatial 174
Gibbs distributions 64
GIS 261, 278
gold standard 215
gradient *see* image gradient
Graham's scan (convex hull algorithm) 92
greylevel 75
   histogram 75
   see also mathematical morphology
ground truth 141, 215

histogram
   bimodal 43
   equalisation 41
   histogram-based transform 41
histological images 212
homogenous coordinates 174, 275
Hough transform 83–4
hull *see* convex hull

ICP *see* iterative closest point
illumination
   diascopic 190
   episcopic 190
   see also lamps;
      shade correction
image
   coordinate frame 230
   pose 229, 230, 231
   primitives 225

Index

image acquisition 1, 161
  software 162
  see also camera
image analysis 69–109, 111
image comparison 264
image domain 263
image enhancement 37–67, 99
  wavelet 100
image fusion 197
image gradient 76
  thresholding 76
  see also edge
image invariants 180–181
  orientation 181
  proportion 181
  translation 180
image matching 167
image moments 180
image processing 37–67
  bottom-up 223
  top-down 223
image registration 167
  see also alignment
image re-sampling 269
image restoration 29
image segmentation 70, 72–85
  boundary approach 72
  edge approach 72
  Kirsch's method 76
  region approach 72
  thresholding 73–75
image sharpening 48
image similarity 184–186
  statistical measures 208–209
    sum of intensity differences 208
    normalised cross-correlation 208
    normalised standard deviation 209
    variance of intensity ratios 208
  information-theoretic measures 209–211
    entropy 209–10
    Shannon-Wiener 210
    mutual information 210
image smoothing 47, 53, 78
image transform
  functional intensity transform 40
  Harr 101
  histogram-based transform 41
  hit-or-miss 87
  linear 46
  morphological 53–60
  non-linear 50
  texture 65
  wavelet 52
  see also filters
image warping 261
Image-Pro Plus 106
ImageTool 105
immersion medium 15

interphase nucleus 2, 29
interpolation 269
  bilinear 99, 175, 269, 270
  nearest neighbour 175, 269, 270
  sinc function 269
  spline 175
  see also thin-plate spline; transformations
ISODATA algorithm 129, 130
isosensitivity curves 146
iterative closest point (ICP) 204–207

jackknife method 141, 142
  see also classifier validation
Jarvis march (convex hull algorithm) 92–93

Karhunen-Loève expansion 128
karyotype 69, 71, 99
  ideogram 99
  karyotyping 113
KBVision 106
kernel methods 121
Kirsch edge operator 80
k-means algorithm 129, 134
knee 225, 229
Köhler, August 7
  illumination 8
Kohonen *see* neural network

lamps
  arc 14
  halogen 14
  tungsten 14
landmark points 224, 226, 227–228
landmarks anatomical 198, 203
Laplacian operator 77
  Laplacian of Gaussian 78
  see also edge
laser 15
leave-one-out method 142, 231, 232
  see also classifier validation
lenses 9–12
  condenser 8, 17
  eyepiece 17
  objective 8, 9, 15
    achromatic 15
    fluorite 15
    semi-apochromat 15
    apochromat 15
light
  absorption of 6
  coherent 6

Index

colour of 3
diffraction of 3–4
laser 15
monochromatic 7
reflection of 5
refraction of 6
scatter of 5
sources 14
likelihood 114
LView Pro 105

magnetic resonance imaging (MRI) 154, 197, 224, 241
microscopy 155
Mahalanobis *see* distance
Markov random field 64
mathematical morphology 37, 53–60, 88–91
  closing 54, 89
  dilation 54, 89
  erosion 54, 89
  greylevel 56
  max and min filters 56
  opening 54, 89
  rolling ball transform 57
  skeletonisation 90–91
  thinning 90
  top-hat transform 58
max filter 56
  see also mathematical morphology
medial axis 276
metaphase
  cell 70, 71
  finding 70, 104
microscope 1–36
  confocal 13
  fluorescence 2, 13
  illumination 14
  image 1, 69
microtome 156–157, 190
  horizontal 190
  sections 159, 261
min filter 56
  see also mathematical morphology
mitosis 70
model
  active shape 234
  coordinate frame 230
  deformable 228
  grey-level 235
  image 175
  matching 232
  parameters 226, 230
  shape 228, 230
  statistical 235
model-based
  interpolation 262
  methods 223–247

modes of variation 238
moments 250
morphing *see* image warping
morphological operators *see* mathematical morphology
morphometric analysis 225
multiresolution
  active shape models 236
  edge detection 81
  gaussian 235
  pyramid 236
  search 237–238
  wavelets 100
mutual information 276
mycelia, fungal 37, 38

neural networks 131–137
  artificial neural networks 131
  competitive network 135
  feedforward 132
  Kohonen self-organising map 136
  multi-layer perception 132
    training 133–134
  radial basis function (RBF) 134
  supervised training 132–135
  training 132
    convergence 133
    epoch 133
    learning rate 133
    momentum 133
    winner-take all 135, 136
  unsupervised training 135–137
NIH Image 105
noise 25
  in cameras 25–26
    blooming 26
    dark current 25
    dead pixels 27
    photon noise 25
    quantisation noise 26
    readout noise 26
    smearing 26
non-parametric classifier 96
normal distribution 115
numerical aperture 3, 9–11
Nyquist frequency 25

opening *see* mathematical morphology
operating point (ROC analysis) 146
optical distortions 11
optical filters 17
  bandpass 18
  colour 18
  contrast 17

Index

optical filters (*continued*)
  dichroic 18
  emission 18
  excitation 18
  infrared, heat absorbing 18
  interference 18
  neutral density 18
  UV-absorbing 19
  wedge 18
optimisation 232
  genetic algorithm 232
  hill-climbing 104
  Levenberg-Marquardt 212
  Powell's method 232, 234
  Simplex method 232, 234
  simulated annealing 234
  see also search
OR *see* binary image processing

pattern recognition 111–152
peaking of classifier performance 138
perceptron *see* neural network
perspective transformation 261
PET *see* positron emmision tomography
photons 3
pin-cushion distortion 12
plane curve
  estimating length 253
point spread function 10
point-set 263
polyline
  mapping open 280
  mapping closed 282
pose 229
positron emmision tomography (PET) 197, 204
Prewit edge operator *see* edge
principal components 228
  analysis 128, 244–6
  eigenvalues 244
  eigenvectors 244
  modes 229
probability
  class-conditional 114
  posterior 115
  prior 114
  see also likelihood
probes, labelled 252
Procrustes transform 242, 277
projection 250
projective stereology *see* stereology
pseudocolour 42

radial basis function *see* neural networks

random variable 250
  3D projection 250
RBF *see* neural networks
receiver operating characteristic (ROC) 112, 145–8
  free-response (FROC) 147
  regular 147
reference set 123
region growing 84–85
region of interest (ROI) 33
registration 197–9, 204–218
  3D–3D 212–15
  contour-based 204
  surface-based 204
  validation 215
  voxel-based 207–211
  see also alignment
registration transformation 199–202
  affine 200
  elastic 201, 202
  projective 201
  rigid 200
resolution 9–10
  limit of microscope 9
  spatial 166
resubstitution 141
RMS error 98
Roberts edge operator *see* edge
ROC *see* receiver operating characteristic
ROI *see* region of interest
rolling-ball *see* mathematical morphology
rotation 182
  ambiguity 184

sample preparation 29, 37
sampling 25, 236, 269
scale
  estimation 182
  and edge detection 51
  in wavelet transform 52
scan-line processing 271
search
  heuristic, for edge linking 81
  see also optimisation
sections
  serial 153, 158
  see also microtome;
    three-dimensional reconstruction
shade correction 29, 59
shear transformation 201
signal to noise ratio (SNR) 20
simulated annealing 102, 128
single-photon emmision tomography (SPECT) 197
singular value decomposition 178, 202, 263
skeleton 90–91

Index

grassfire technique 90
pruning 90
thickening 90
for touch and overlap resolution 92
see also mathematical morphology
Skil-Image 105
smoothing 225, 236
see also image smoothing
snake *see* active contour
SNR *see* signal to noise ratio
sobel edge detection *see* edge
spatial data mapping 261
speckle 7
SPECT *see* single-photon emmission tomography
spherical aberration 12, 261
spline 262
spot counting 29–35
stepwise selection of features 126
stereology 155, 249–60
   3D estimation 249
   projective 249–60
stereotactic frame 199
structure intersections 159
sub-sampling 236
see also sampling
surface areas
   estimating 254–5
surgery 198

Talairach space 198
texture 37, 60–66, 233
   co-occurrence matrix 61
   Gibbs models 64
   greylevel difference statistics 63
   Markov random field 64
   measuring 65
   modelling approach 60, 64
   run-length statistics 63
   statistical approach 61–63
   structural approach 60, 64
   texture transforms 65
therapy planning 198
thinning 90
see also mathematical morphology
thin-plate spline 202, 270, 282–63
three-dimensional
   analysis 196
   measurement 155
   reconstruction 153–195
   registration 196–222
thresholding 40, 73–75
   adaptive 73–75

for classification 116
colour 103
fixed 73
value selection 45, 75
tie-points 262
TIM for Windows 105
top-hat transform 58
see also mathematical morphology
touch and overlap resolution 92–93
transformations
   affine 177, 262, 274, 275–278
   conformal 274, 279–282
   global 274
   gaussian 284
   polynomial 274, 278–279
   multiquadric 284
   see also thin-plate spline
translation 182
transportation algorithm 102
triangulation 271
true positive, negative 144
tumour 204
two-class problem 144

variance 250
vignetting 58
virtual sectioning 154
Visible Human 217
Visilog 106
volumes 255
voxel
   similarity 208
   size 203

watershed algorithm 76, 77
wavelets 52, 100–101
   Harr transform 101
wax model (3D) 153
weighted density distribution (WDD) 95
Wilks' lambda 127
WiT 106
world image 175

xmorph 286
X-ray CT 197, 204

zooming 46